SpringerBriefs in Geography

More information about this series at http://www.springer.com/series/10050

Andrew Emmanuel Okem
Editor

Theoretical and Empirical Studies on Cooperatives

Lessons for Cooperatives in South Africa

Springer

Editor
Andrew Emmanuel Okem
Durban
South Africa

ISSN 2211-4165 ISSN 2211-4173 (electronic)
SpringerBriefs in Geography
ISBN 978-3-319-34215-3 ISBN 978-3-319-34216-0 (eBook)
DOI 10.1007/978-3-319-34216-0

Library of Congress Control Number: 2016940805

Printed on acid-free paper

This Springer imprint is published by Springer Nature
The registered company is Springer International Publishing AG Switzerland

Contents

Introduction and Book Overview

Modern cooperatives have been instrumental in uplifting the social and economic conditions of people around the world since their emergence in England in 1844. As an economic model that seeks to advance members' collective interests, the cooperative movement has come to be viewed as an invaluable tool for human and economic development. Over the years, the cooperative movement has transformed from being an economic model mainly for the poor and vulnerable to becoming a significant actor in the global economy. Despite the growing significance of the cooperative model, a number of challenges continue to preclude cooperatives in South Africa from attaining their esteemed socio-economic ideals. Against the backdrop of the increasing global recognition of the roles of cooperatives in socio-economic development, this book seeks to present selected studies on the cooperative movement and draw important lessons for South Africa. This book comprises both theoretical and empirical works.

The chapters were selected on the basis that they provide valuable lessons for the South African cooperative sector. The theoretical chapters in the volume engage debates on the meaning, values relevance, and application of the concept of cooperatives while empirical chapters present pertinent lessons about factors that facilitate the success of cooperatives. There are two chapters on the South African cooperative sector in the book. The first chapter presents a history of the country's cooperative sector and attempt to demonstrate how the country's political system has and continues to influence the cooperative sector. The second chapter presents a case study of a secondary cooperative and outlines the constraints faced by the country's cooperatives. It is expected that contributions in this book will translate into novel analytical frameworks, conceptual approaches, and empirical accounts of how the cooperative movement in South Africa can be strengthened. Resources contained in the volume are valuable for academics, policymakers, development practitioners, students and members of cooperatives.

This book is organised into nine chapters. Chapter 1 presents an overview of the cooperative movement. This chapter discusses issues related to the conceptualisation of cooperatives and the principles and values of the cooperative movement.

This chapter notes that these principles and values are what define and make cooperatives a unique organisational form for addressing people's socio-economic needs.

Chapter 1 examines the meaning, defining characteristics, and forms of cooperatives. This chapter argues that these very same characteristics make cooperatives powerful vehicles for socio-economic empowerment that in turn contribute to local economic development. The review shows that although cooperatives emerged as small-scale bottom-up organisations to addressing prevailing socio-economic conditions during the industrial revolution, they have evolved over the years and now make important contributions to the global economy.

In Chap. 2, Okem and Stanton argue that although modern cooperatives were introduced in Africa during colonialism, cooperatives in Africa predate the colonial era. Pre-colonial form of cooperative, according to the authors, can still be found today despite the predominance of modern forms of cooperatives. From the historical overview of the evolution of cooperatives in Africa, the authors established that cooperatives are more successful in contexts where they operate independently of government interference. For this reason, the authors recommend that the role of government should be limited only to those that pertain to the creation of conditions for the emergence and growth of cooperatives.

In Chap. 3, Fakude engages debates about emerging interdisciplinary perspectives on the cooperative movement. The interdisciplinary themes explored in the chapter include cooperatives' role in Africa's economic growth and development (Development Economics/Development Studies); the role of social capital in the growth and development of third generation cooperatives (Sociology); and women's participation in cooperatives (Gender Studies). Fakude argues that even though it may be tempting to look for theoretical debates in the emerging interdisciplinary themes in relation to cooperatives, it should be acknowledged that the themes are merely differing perspectives which may or may not develop into full-fledged debates as more academic interest is generated in the sector. The chapter concludes by pointing out that the areas of interests raised in the chapter require further research in order to ground the debates.

Chapter 4 assessed the roles of cooperatives in post-conflict societies. Ettang and Okem argue that protracted conflicts have existed in Africa for decades and have had a negative impact on socio-economic growth and stability. The authors examined how cooperatives are useful in post-conflict environments and the challenges and limitations they face in fragile and post-conflict societies. Ettang and Okem argue that in limiting future occurrence of violent conflict, cooperatives bring forth linkages between development and peacebuilding (two very important factors in preventing future conflict and guaranteeing sustainable peace). Recognising this role of cooperatives in post-conflict reconstruction, the authors recommend that more attention be paid to cooperatives and their benefits in post-conflict environments particularly in Africa. Ettang and Okem further argue that cooperatives should be adopted as a key component of Poverty Reduction Strategy Programmes that offer short, immediate and long-term responses in addressing the root causes of conflict.

In Chap. 5, Okem and Tshishonga examine the Kenyan cooperative sector and attempt to draw lessons for cooperatives in other contexts. The authors argue that although cooperatives were introduced during the colonial era and adopted by the post-colonial government, it was only after cooperatives were made independent of the state did they begin to experience some measures of success. The chapter notes that applying the Kenyan model is critical to the success of cooperatives in other African contexts.

Chapter 6 examines factors that contributed to the collapse of coffee cooperatives in the post-adjustment period in Cameroon. This chapter began by examining the strategic roles of cooperatives in the coffee value chain prior to the introduction of the structural adjustment programme. The author argues that the failure of coffee cooperatives in the post-adjustment era is linked to the withdrawal of state control of the sector that opened it up to stiff competition which exposed the inefficiencies of coffee cooperatives.

Okbandrias and Okem provide a historical overview of the cooperative movement in South Africa in Chap. 7. The point of departure of the chapter is whether the current interest in cooperatives in South Africa can be interpreted as a reflection of genuine interest in the cooperative movement or a mere demonstration of a trend that has gained prominence with no entrenchment of the principles and values of the cooperative movement. Through a review of the country's cooperative sector, Okbandrias and Okem argue that the current approach to cooperatives paints a picture that cooperatives are not deeply entrenched. The primary argument here is that the country is characterised by a paradox of a high number of newly registered cooperatives and high failure. Okbandrias and Okem interpret this paradox as a reflection that government financial incentive is creating an environment that fosters the emergence of unsustainable cooperatives. This chapter concludes with an argument that cooperatives in South Africa are nothing but a trend that fails to capture the rich values they ought to espouse.

Chapter 8 examines how the Inanda Farmer's Association contributes to the development and sustainability of cooperatives in the Inanda area. This chapter evaluated the association in terms of its challenges and strengths in providing institutional support aimed at the development and sustainability of member agricultural cooperatives. Tshishonga and Bandyambona argue that the failure of the association in dealing with challenges such as poverty, unemployment and inequality is due to the lack of understanding of the principles of the cooperative movement. Findings of this chapter demonstrated that for cooperatives in Inanda to be sustainable, cooperatives themselves should take on the primary responsibility of addressing their problems rather than relying primarily on government support.

In Chap. 9, Okem and Tshishonga present key lessons that could be drawn from preceding chapters and applied to the South African context. The lessons drawn from the chapters presented in this book include the need for cooperatives to claim their unique identity as a social movement, ensure a radical application of the principles and values of the cooperative movement and utilising the strength of network and supply chain. Other lessons pointed out in the chapter include uprooting dependency through self-help, furthering the education and training of cooperatives and extending cooperatives beyond focusing on only poor and vulnerable communities.

Andrew Emmanuel Okem
School of Built Environment and Development Studies
University of KwaZulu-Natal

Chapter 1
The Meaning and Defining Characteristics of Cooperatives

Andrew Emmanuel Okem

Abstract Around the world, cooperatives are increasingly being recognised for their invaluable roles in socio-economic development. This recognition has translated into countries and development organisations taking proactive steps towards grounding the cooperative movement so as to maximise their socio-economic benefits. The primary objective of this chapter is to present an overview of the cooperative movement. In doing this, the chapter engages how the cooperative movement has been conceptualised. The chapter also presents the defining characteristics of the cooperative movement and how these translate into the impact they have on the society. The chapter notes that the unique characteristics of cooperatives are what make the sector important in addressing socio-economic challenges.

Keywords Cooperative · Development · Employment · Principles · Values

1.1 Introduction and Chapter Overview

Over the years, the cooperative movement has played an important role in addressing the challenges of poverty, unemployment and building social capital. Despite the stated benefits of the cooperative movement, its application has been associated with confusion in some contexts owing to poor understanding of what a cooperative is and how it functions. This chapter presents an overview of the cooperative movement. The chapter is organized into three sections. In Sect. 1.2, an overview of how cooperatives are conceptualised is presented. A review of the principles and values of the cooperative movement is also included in Sect. 1.2. The review discusses how the principles and values define the character of the cooperative movement. In Sect. 1.3, the different forms of cooperatives are

A.E. Okem (✉)
School of Built Environment and Development Studies, University of KwaZulu-Natal, Durban, South Africa
e-mail: andokem@gmail.com; okem@ukzn.ac.za

A.E. Okem (ed.), *Theoretical and Empirical Studies on Cooperatives*,
SpringerBriefs in Geography, DOI 10.1007/978-3-319-34216-0_1

presented. The section discusses the characteristics of the different forms of cooperatives and benefits associated with each form. The last section of the chapter presents the contributions of cooperatives to the global economy. The goal of this section is to challenge the notion that cooperatives are only applicable in addressing the economic conditions of vulnerable populations in developing countries. The review demonstrates that cooperatives, although have and continue to play a vital part in alleviating the poor socio-economic conditions of the vulnerable population, need not to be limited to that population.

1.2 An Overview of the Conceptual Understanding of Cooperatives

Literature shows there are different definitions of a cooperative. Porter and Scully (1987: 494) define cooperatives as "voluntary closed organizations in which the decision-control and risk-bearing functions repose in the membership, and decision management responses in the agent (manager), who represents the principal's interests". This definition presents three characteristics of a cooperative. Firstly, a cooperative is a voluntary association of persons. Such an association is formed to address members' common needs. Secondly, a cooperative is controlled by its members since they are actively involved in making decisions about its operations. The third feature of the definition points to the fact that the risks of a cooperative are borne by members since it is member-owned and member-controlled.

According to Torgerson et al. (1997: 2), cooperatives could be seen "as a social movement of independent farm operators seeking to enhance and protect their place in the economic organization of agriculture". The overriding aim of a cooperative, according to this view, is to protect the interest of members against exploitation by actors such as middlemen in the supply chain. The foregoing suggests cooperatives exist to balance the market economy by countering the effects of market distortion such as monopoly, monopsony, and oligopoly (Baker and Graber-Lüzhøft 2007). However, a weakness in Torgerson, Reynolds, and Gray's conception of cooperatives is that it limits cooperatives to the agricultural sector. This is often a feature in the conceptualisation of cooperatives as agricultural organisations ideal for rural peasants. This is rather a misleading conception. As I point out later in this chapter, cooperatives are neither limited to the agricultural sector nor are they suitable only for the rural poor.

Cooperatives bring individuals together to empower themselves through the power of the collective. This value of cooperatives was acknowledged in a United Nations (1996) report which views cooperatives as an effective tool that enables a group of people (whether small or large) to mobilise resources to meet their needs. The report noted that in the process of resource mobilisation to meet members' needs, cooperatives foster entrepreneurship in communities where they operate. In this way, cooperatives improve the socio-economic condition of members as well as that of the community. What is unique about a cooperative is that those who own a cooperative are often its customers (Birchall and Ketilson 2009). This characteristic sets cooperatives apart from other forms of businesses such as Investor-Owned Firms (IOFs) (Mazzarol 2009). Unlike IOFs in which members only contribute capital, members of a cooperative contribute capital and utilise services or market provided by the cooperative (Dlamini 2010).

Although the foregoing views show that cooperatives have been conceptualised in a number of ways (each with subtle differences), there has been a growing acceptance of the definition of cooperatives by the International Cooperative Alliance (ICA).[1] The ICA defines a cooperative as "an autonomous association of persons united voluntarily to meet their common economic, social, and cultural needs and aspirations through a jointly owned and democratically controlled enterprise organised and operated on cooperative principles" (International Cooperative Alliance 1995). The strength of ICA's definition lies in the fact that it synthesizes the diverse definitions of cooperatives in the literature. In addition, the definition does not only focus on economic values, it also emphasises the social and cultural values of the cooperative movement. This aspect is often absent in other definitions of cooperatives. The definition of cooperatives proposed by the ICA is what underpins the understanding of cooperatives in this book.

In line with ICA's definition, there is an increasing emphasis on non-economic benefits of cooperatives. What is important here is the recognition that self-organisation through a cooperative is an important strategy that simultaneously achieves social and economic outcomes. The combination of associationism and market forces makes cooperatives a veritable market option. Against this backdrop, Menzani and Zamagni (2010: 21) argue that "cooperatives can be seen as an instrument for unleashing the market's full potential" primarily because of their economic and social outlook. In the absence of this recognition, the successes and failures of cooperatives would be defined in purely economic terms at the expense of the social and cultural dimensions (Garnevska et al. 2011; Mellor 2009). In recognition of the non-economic benefits of cooperatives, Wanyama et al. (2009: 187) emphasise the importance of adopting a "holistic social economy model" when it comes to the study of cooperatives. This approach is particularly important in light of the principles and values that underpin the cooperative movement.

[1]Headquartered in Geneva, Switzerland, the ICA is the international umbrella for the global cooperative movement. Its primary goal is to unite the world's cooperative movement by providing a platform for cooperatives to come together to network, support and share knowledge. It sees itself as the custodian of the principles and values of the cooperative movement. The ICA provides various supports to the cooperative movement including advocacy, capacity building and knowledge sharing.

1.3 Principles and Values of the Cooperative Movement

Across the world, the cooperative movement is guided by seven key principles. By adhering to the principles, the identity of cooperatives is upheld regardless of where they are located. Each of the seven principles will be briefly explored in the following paragraphs. This book recognises the conceptual value of these seven principles insofar as they reiterate the fundamental rationale for cooperatives, as well as their potential strength.

1.3.1 Voluntary and Open Membership

The first underlying principle of the cooperative movement is that of voluntary and open membership (International Cooperative Alliance 2007). In realising this principle, cooperatives do not discriminate on the basis of gender, social, racial, political or religious status of people. According to the principle of voluntary and open membership, a cooperative accepts anyone intending to utilise the services or opportunities provided by a cooperative. However, such a person must subscribe to the values for which the cooperative was established. Furthermore, a prospective member must be ready to accept the responsibilities that come with being a member of the cooperative. Such responsibilities include contributing time, finance, inputs and other resources required for the functioning of the cooperative. In addition, members are responsible for the governance of cooperatives since a cooperative is a member-owned organisation.

1.3.2 Democratic Member Control

Apart from being committed to open and voluntary membership, cooperatives are subject to member control (Nilsson et al. 2012). The principle of democratic member control is at the core of the cooperative movement. This principle ensures that at every point, a cooperative is committed to fulfilling the needs and aspirations of those who own it. The control of cooperatives is realised through the "one-member-one-vote" strategy (Bernard and Spielman 2009: 61). This principle safeguards against the hijacking of a cooperative either by members with higher investment or non-cooperative institutions (such as government and development organisations). Given that a cooperative is owned and controlled by members, each member is entitled to benefit from its services and profits. In view of this, the International Cooperative Alliance (2007) notes that surpluses of cooperatives are utilised by members for a number of reasons including the improvement of their cooperative and direct payouts from cooperatives proportionate to their contributions. In addition, surpluses of cooperatives are used in other activities such as community development programmes that are supported by members.

1.3.3 Member Economic Participation

Member economic participation is anchored on the view that cooperatives are owned by members who are required to contribute equally to its growth (International Cooperative Alliance 2007). The nature of contributions expected of members is dependent on the kind of cooperative and the sector in which it operates. For this reason, member economic participation could be in form of buying shares in a cooperative, delivering farm produce to a cooperative-owned processing plant, contributing manual labour to an agricultural cooperative and so on.

1.3.4 Autonomy and Independence

Cooperatives are autonomous and independent organisations. Adherence to this principle guarantees that cooperatives are free from the influence of external parties such as government, non-governmental organisations (NGOs), and development agencies (Nannyonjo 2013). Autonomy guarantees that cooperatives are not manipulated into fulfilling the needs and aspirations of external agents at the expense of members' needs and aspirations (Hannan 2014). Abiding by this principle entails that government should focus on creating enabling environments that will foster the growth of cooperatives as autonomous organisations. Such enabling environment includes the implementation of favourable tax policies that encourage ploughing profits back into cooperatives thus strengthening the cooperative movement.

1.3.5 Education, Training, and Information

Education, training, and information is the fifth principle of the cooperative movement (International Cooperative Alliance 2007). According to this principle, a cooperative ought to provide relevant education and training to its members (Wanyama et al. 2009). Education and training empowers and encourages members of cooperatives to participate in decision-making regarding the running of a cooperative. It also guarantees that leaders of a cooperative have the requisite skills to oversee the day-to-day running of the cooperative. Furthermore, this principle commits cooperatives to educating the public about their nature, working and benefits, thus ensuring that the public is informed about the rationale for the formation of and participation in cooperatives. This could result in the public choosing to either become members of an existing cooperative or to establish a new one. Effectively, member education can contribute to increased number of cooperatives and cooperators.

1.3.6 Cooperation Among Cooperatives

Promoting cooperation among cooperatives is another principle of the cooperative movement. According to Novkovic (2008), cooperation among cooperatives facilitates the formation of a network of cooperatives at local, regional, national and international levels. By developing a robust network, cooperatives become strategically placed to harness the benefits of economies of scale and scope (Dredge 2006). Cooperation among cooperatives enhances the viability and sustainability of the cooperative movement since it links them to support and opportunities in the network. Furthermore, cooperation results in the circulation of income and business opportunities among cooperatives. What this entails is that instead of outsourcing activities to other business, business opportunities are retained among cooperatives thus ensuring that cooperatives provide business services to one another. This is particularly important to new and emerging cooperatives that need new businesses and access to resources in order to function.

1.3.7 Concern for the Community

As an economic model that is rooted in the grassroots, cooperatives are expected to have a strong concern for the immediate community in which they operate (International Cooperative Alliance 2007). According to Zeuli et al. (2004: 18), cooperatives have "the potential to create more substantial social and economic benefits within a community than non-cooperative firms". Similarly, Bertulfo (2007: 106) contends that "while focusing on member needs and wishes, cooperatives work for the sustainable development of their communities". The argument here is that addressing members' socio-economic needs will have a spillover effect on the community at large. Benefits that accrue from cooperatives to the community are either intentional or unintentional. While the former relates to purposive and planned action, the latter is concerned with the trickle-down benefits of the activities of cooperatives to local communities. When people participate in cooperatives, they advance their personal, collective and communal interests.

In addition to the principles outlined above, cooperatives are governed by a set of values. According to the International Cooperative Alliance (2007), the cooperative movement is underpinned by the "values of self-help, self-responsibility, democracy, equality, equity and solidarity". These values ensure that cooperatives are not only geared towards the economic good of members but are also about empowering members to become better agents in their communities. Cooperatives are also governed by the ethical values of honesty, openness, social responsibility and caring for others (International Cooperative Alliance 2007). Such values are expected to be the essence of every cooperative.

What the foregoing shows is that, essentially a cooperative exists to address the socio-economic needs of its members. The hypothesis is that values of self-help, self-responsibility, democracy, equality, equity, and solidarity make cooperatives ideal for socio-economic development. Their developmental orientation is particularly valuable in developing regions like sub-Saharan Africa where almost 80 % of the population are involved in subsistence activities such as small-scale agriculture (Nyiraneze 2007). Studies (see, for instance, Develtere et al. 2008; Getnet and Anullo 2012) have shown that cooperatives have been instrumental in poverty alleviation, job creation, economic development and social transformation.

According to Mazzarol (2009), the benefits of cooperatives depart markedly from regular businesses activities that are underpinned by the philosophy of competition and profit maximisation. Due to its focus on members' needs, the cooperative model provides an alternative model for the economic upliftment of disadvantaged people. The cooperative model, particularly agricultural cooperatives, provides a parallel supply chain network for its members to circumvent the negative externalities created by intermediaries in the agricultural supply chain. The principles and values of the cooperative movement presented above underpin the different forms of cooperatives.

1.4 Forms of Cooperatives

According to Birchall and Ketilson (2009: 13), cooperatives can be categorised based on the nature of stakeholders in a cooperative enterprise namely consumers, producers, and workers. Using this criterion, Birchall and Ketilson (2009) grouped cooperatives into four main forms: consumer, worker, producer, and finance cooperatives. Each form of cooperative will be briefly explored in the following paragraphs.

A consumer cooperative is a form of cooperative owned by people involved in the purchase of consumable goods. Such cooperatives buy goods in bulk and are therefore able to sell these to members at reduced prices (Deller et al. 2009). Consumer cooperatives can exist as supermarkets, grocery stores and retail outlets that sell different items to members at lower prices (Little et al. 2010). In this sense, their existence is informed by the desire to break the market monopoly by providing parallel supply chains for consumers. In the US, for instance, consumer cooperatives have been construed as "a new hope for the generation of Americans who looked toward a new economic order to distribute the nation's resources more equitably" (Williamson 2005: 61). New cooperatives most frequently were a symbol of rebellion against technocracy, hierarchical corporations, and "big business as usual". According to the University of Wisconsin Center for Cooperatives (2012), William King (1786–1965) advocated and popularised consumer cooperatives. Through his magazine, "The Cooperator", he circulated his message of consumer cooperatives, which resulted in rapid expansion of consumer cooperatives.

Burdin and Dean (2009: 518) define worker cooperative "as an enterprise where the firm's labor force chooses the management and the administrative structure using a democratic political process". Either a small or a large group of people can own a worker cooperative. For instance, a tourism business can be owned by its workers just as a group of factory workers could own the factory in which they work. According to Majee and Hoyt (2010: 147), worker cooperatives, "through open membership and democratic local ownership of the enterprise, are believed to provide a platform on which participation of local people in both social and economic activities can be enhanced". In addition, worker cooperatives, because they are member-owned, have the potential for improved performance. Mathie (2012: 22) notes that "studies show that employee owned businesses tend overall to have higher productivity, greater levels of innovation, better resilience to economic turbulence and more engaged and fulfilled workers who are less stressed than colleagues in conventionally owned businesses". These benefits of worker cooperatives stem from the fact that they strive to assert workers' rights. Consequently, the health and wellbeing of members cannot be sacrificed for profit. This is why Stikkers (2011) argues that the success of worker cooperatives is measured by the extent to which they create quality and sustainable jobs. The success of worker cooperatives has also been linked to the resilience of worker cooperatives in times of economic crisis. For instance, Birchall and Ketilson (2009) argue that despite the effects of the 2008 crisis, worker cooperatives in Italy remained largely functional. At the height of the Spanish financial crisis, cooperatives in the Basque region ensured that unemployment was consistently kept low. Cooperatives achieved this by rehiring those laid off by poorly performing cooperatives (Birchall and Ketilson 2009). The resilience of cooperatives in any given economic crisis is also demonstrated by their ability to create and sustain employment. Smith and Rothbaum (2013: 4) cites some examples including France where "employment in cooperatives grew 12.1 % and in all firms by 1.4 %; in Italy, the figures were 86.2 % for cooperatives compared to 3.8 % for all firms; in Spain, 31 % versus −8.1 %; in the UK, 133 % compared to −2.0 %; and in the EEC as a whole, 76 % compared to 2.0 %". In a study comparing the performance of Capitalist Firms (CFs) and Worker Cooperatives (WCS), Burdin and Dean (2009: 527), found that "CFs would produce a socially inefficient level of layoffs due to their inability to establish credible commitments between owners and workers. By contrast, because of their unique control structure, WCs would have more egalitarian adjustment mechanisms at their disposal".

Worker cooperatives can emerge in one of three ways. The first of these are new start-up cooperatives that emerge in response to needs in a niche market. To take advantage of the recognised niche market, a group of individuals comes together to start a business in which they are simultaneously its workers. The second way that a worker cooperative emerges is through the conversion of an existing business into a cooperative (Burridge 2012). This could happen when the owner of a business sells his/her business because s/he wants to retire or is unable or no longer desires to continue running the business. Worker cooperatives also emerge in times of economic difficulties. In this case, an existing business is rescued by workers to preserve it from liquidation (Artz and Kim 2011; Burridge 2012). By putting together their income to purchase shares in the company, worker cooperatives that emerge in this way achieve three things: they save the company from liquidation; save their jobs, and create a means of obtaining additional income through the profit of the cooperative (Burridge 2012).

A producer cooperative refers to the form of cooperative owned by a group of individuals that produce the same kind of goods. Members of this form of cooperative use shared facilities for processing and distributing their goods. Known sometimes as a marketing cooperative, producer cooperatives provide value added services including processing, packaging, and branding to members. Individuals who join producer cooperatives enjoy services that would otherwise be expensive if sourced individually. Producer cooperatives are often found in the agricultural sector where they provide value added services to members (Bloom and Hinrichs 2011). For instance, corn producers can come together to form a cooperative that owns and operates a flour milling plant. The cooperative could also provide branding services to independent farmers who belong to the cooperative. This gives them access to larger markets as well as reduced transactions costs. In this way, producer cooperatives simultaneously lower input costs as well as increase farmers' bargaining power which could result in higher returns on investment (Mazzarol 2009).

Financial cooperatives are the fourth form of cooperatives. Financial cooperatives, like other forms, are member-owned. Members' savings provide the reserve that financial cooperatives lend to both members and non-members (Birchall 2013). Cooperative financial institutions (CFIs) are better sources of finance for cooperatives since they do not place restrictions that emerging cooperatives face when they source finance from commercial financial institutions.

Studies have shown that CFIs are resilient institutions during economic crises (Crear 2009; Delbono and Reggiani 2013; Birchall 2013). This characteristic of CFIs has been attributed to their commitment to promoting members' needs. A study by Birchall (2013) shows that the resilience of CFIs is linked to their values of concern for members and the community. Their values make cooperatives engage in banking practices that do not expose members' finance to undue risks. Although the 2008 financial crisis had negative effects on the CFIs, they largely performed better than other kinds of financial institutions (Crear 2009; Delbono and Reggiani 2013). According to Birchall (2013: 2), most financial cooperatives survived the 2008 financial crises "without needing any government bailouts, without ceasing to lend to individuals and businesses, and with the admiration of a growing number of people disillusioned with 'casino capitalism". Similarly, a study by the International Cooperative Alliance (2010) found that "cooperative banks [since the 2008 recession] gave millions of people stability and financial security because the cooperative banking business model emphasises not profit maximisation but instead the best possible products and services to members". Birchall and Ketilson (2009: 13) further argue that this approach of cooperative banks "show that there is an alternative to the current policy of greater public regulation of private banks, while in many countries also providing banking and insurance to low-income people who would otherwise be unbanked". Effectively, cooperatives have been able to provide financial services to their customers despite the financial crises.

Table 1.1 Summary of the forms of cooperatives

Form	Characteristics	Benefits	Examples
Consumer cooperative	Sell goods/services to members	Members pay for goods and services at reduced price	supermarkets, grocery stores, retail outlets
Worker cooperative	Owners are simultaneously its employees	Guarantees employment of workers; improves worker commitment, working condition, wages, and productivity	Tourism business, cleaning agencies, farms, construction company
Producer cooperative	Found mainly in the agricultural sector and provides value added services to members	Reduces transaction costs through the use of shared facilities, common marketing outlets, and common brand	Shared milling plant, shared marketing outlet
Financial cooperative	Operates in the financial sector. Its capital is sourced from members' contributions	Resilient financial institutions; better source of finance for cooperatives	Cooperative banks, savings and credit cooperative societies, insurance

Source Author's creation

1.4.1 Summary of the Forms of Cooperatives

Table 1.1 gives a summary of the forms of cooperatives discussed above. It is important to point out that the different forms of cooperatives explored above have similar organisational forms and are guided by the same principles and values of the cooperative movement. The different forms of cooperatives are created primarily to meet members' socio-economic needs. These needs could be access to marketing opportunities, housing, credit facilities, urban renewal, use of shared facility, meeting social needs and so on.

1.5 The Contributions of Cooperatives to the Global Economy

The different forms of cooperatives summarised above make important contributions to the global economy. According to Merrien (2014), there are about 2.6 million cooperatives around the world. These cooperatives "have over 1 Billion memberships and clients" with "12.6 Million Employees work[ing] in 770,000 Cooperative offices and Outlets" (Merrien 2014: 1). Cooperatives generate an annual revenue to the tune of US$3 trillion (Merrien 2014: 1). The cooperative sector also has a strong presence in the global insurance sector. According to the

Table 1.2 Countries by total turnover of cooperatives over 100 Million US$

Countries	Turnover (billion US$)	Countries	Turnover (billion US$)
USA	662.23	New Zealand	30.22
France	363.63	Norway	29.07
Japan	358.81	Belgium	23.38
Germany	284.08	Sweden	21.12
Netherlands	116.23	Australia	19.14
Italy	95.06	Singapore	5.30
Spain	85.61	Ireland	5.20
Switzerland	85.51	India	4.41
UK	84.15	Colombia	3.23
Finland	64.11	Argentina	1.96
Canada	52.33	Portugal	1.91
Denmark	51.64	Malaysia	1.70
South Korea	39.35	Saudi Arabia	1.18
Austria	31.39	Other countries	4.50
Brazil	30.30		

Source International Cooperative Alliance (2007: 14)

International Cooperative and Mutual Insurance Federation (2014), "the mutual and cooperative sector represented 27.3 % of the global insurance market in 2013, a significant growth in global market share from 23.8 % in 2007". The foregoing underscores the significant roles of cooperatives in the global economy.

Cooperatives have and continue to make valuable contributions to the economies of many high-income countries (see Table 1.2). The economies of these countries are characterised by strong, vibrant and diversified cooperatives that employ large workforce and contribute to the countries' GDP (International Cooperative Alliance 2010). As shown in Table 1.2, cooperatives contributed US$158.75 billion to the global economy in 2013.

The high concentration of cooperatives, coupled with their success in industrialised countries, is bringing about a paradigm shift in the way they are conceptualised and operationalised in these countries. Rather than being seen as an economic model suitable only for vulnerable peasants, scholars and policy makers are beginning to emphasise the need to envision and approach the cooperative movement as a viable economic model for all (International Cooperative Alliance 2010). Consequently, a number of countries are creating enabling environments for cooperatives through various policy instruments that could facilitate the emergence and growth of the cooperative sector (Mutuo 2012). The contrary is the case for developing countries like South Africa where cooperatives are often small-sized and operate mainly in agriculture.

1.6 Conclusion

The primary objective of this chapter was to provide an overview of the meaning and defining characteristics of the cooperative movement. The chapter notes that although there are different definitions of cooperatives, there is a growing recognition and acceptance of the definition offered by the ICA. As noted in the chapter, the value of the definition lies in the fact that it recognises cooperatives as a unique social organisation that works towards meeting members' economic, social and cultural needs. To achieve these goals, cooperatives are underpinned by sets of universal principles and values that set them apart from other forms of business organisations. Cooperatives, as noted in the chapter, have and continue to be an instrument in employment creation, poverty alleviation, and social development. Additionally, they are contributors to the global economy. What the foregoing demonstrates is that cooperatives have the potential to improve the socio-economic conditions of developing countries such as South Africa where there is a high level of poverty, unemployment, economic inequality and social fracture.

References

Artz, G., & Kim, Y. (2011). Business ownership by workers: Are worker cooperatives a viable option? Iowa State University: Department of Economics.

Baker, D., & Graber-Lüzhøft, K. (2007). Concentration in agribusiness and marketing A case of Study of Arla foods. In P. Pinstrup-Andersen & F. Cheng (Eds.), *Food policy for developing countries: The role of government in the global food system* (Vol. 6). Ithaca: Cornell University.

Bernard, T., & Spielman, D. J. (2009). Reaching the rural poor through pural producer organizations? A study of agricultural marketing cooperatives in Ethiopia. *Food Policy, 34*(1), 60–69.

Bertulfo, L.T.Y. Women, Co-op and Peace Work, in Emmanuel, J. and Macpherson, I. (2007). Co-operatives and the Pursuit of Peace, 105–112. Victoria: British Columbia Institute for Co-operative Studies.

Birchall, J. (2013). *Resilience in a downturn: The power of financial cooperatives*. Geneva: International Labour Office.

Birchall, A., & Ketilson, L. (2009). *Resilience of the cooperative business model in times of crisis sustainable enterprise programme*. International Labour Organisation.

Bloom, J. D., & Hinrichs, C. C. (2011). Moving local food through conventional food system infrastructure: Value chain framework comparisons and insights. *Renewable Agriculture and Food Systems, 26*(01), 13–23.

Burdin, G., & Dean, A. (2009). New evidence on wages and employment in worker cooperatives compared with capitalist firms. *Journal of Comparative Economics, 37*(4), 517–533.

Burridge, T. (2012). Basque co-operative Mondragon defies Spain slump. Retrieved 13 January, 2013, from http://www.bbc.co.uk/news/world-europe-19213425

Crear, S. (2009). *Cooperative banks, credit unions and the financial crisis*. Paper presented at the Paper for the UN Expert Group Meeting on Co-operatives, April, New York.

Delbono, F., & Reggiani, C. (2013). Cooperative firms and the crisis: evidence from some Italian mixed oligopolies. *Annals of Public and Cooperative Economics, 84*(4), 383–397.

Deller, S., Hoyt, A., Hueth, B., & Sundaram-Stukel, R. (2009). *Research on the economic impact of cooperatives*: University of Wisconsin Center for Cooperatives.

Develtere, P., Pollet, I., and Wanyama, F. (Eds.). (2008). Cooperating out of Poverty The Renaissance of the African Cooperative Movement. Geneva: International Labour Office.

Dlamini, T. R. (2010). *Collective farming: Elements constituting an effective agricultural co-operative, the case of three co-operatives in the uMgungundlovu district.* (Master of Agriculture (Food Security)), University of KwaZulu-Natal.

Dredge, D. (2006). Policy network and the local organization of tourism. *Tourism Management, 27*, 269–280.

Garnevska, E., Liu, G., & Shadbolt, N. M. (2011). Factors for successful development of farmer cooperatives in Northwest China. *Supporters and Partners, 14*(4), 69.

Getnet, K., & Anullo, T. (2012). Agricultural cooperatives and rural livelihoods: Evidence from Ethiopia. *Annals of Public and Cooperative Economics, 83*(2), 181–198. doi:10.1111/j.1467-8292.2012.00460.x.

Hannan, R. (2014). The institution of co-operation: A pathway from governance to spillover and poverty reduction. *Journal of Co-operative Organization and Management, 2*(1), 34–42. doi:10.1016/j.jcom.2014.03.003.

International Cooperative Alliance. (1995, 26 May 2007). Statement on the co-operative identity. 15 March, from http://www.ica.coop/coop/principles.html

International Cooperative Alliance. (2007). Cooperative principles and values: International cooperative alliance principles and values. Retrieved March 6, 2013, from http://www.cdi.coop/icaprinciples.html

International Cooperative Alliance. (2010). *Global 300 report 2010 international co-operative alliance: The world's major co-operatives and mutual businesses*. Geneva: International Cooperative Alliance.

International Cooperative and Mutual Insurance Federation. (2014). Global mutual and cooperative market infographic 2014. Retrieved February 13, 2015, from http://www.icmif.org/global-mutual-and-cooperative-market-infographic-2014

Little, R., Maye, D., & Ilbery, B. (2010). Collective purchase: Moving local and organic foods beyond the niche market. *Environment and Planning. A, 42*(8), 1797.

Majee, W. & Hoyt, A. (2010). Are Worker-owned Cooperatives the Brewing Pots for Social Capital? Community Development, *41*(4), 417–430.

Mathie, J. (2012). *Co-operative farming is vital for the UK—not just for rural communities but the security of the country as a whole*. Borehamwood: Mutuo.

Mazzarol, T. (2009). Cooperative enterprise: A discussion paper and literature review. Retrieved January 14, 2013, from http://www.cemi.com.au/sites/all/publications/0901%20COOPS%20WA%20LIT%20REVIEW.pdf

Mellor, J. W. (2009). Measuring Cooperative Success: New Challenges and Opportunities in Low- and Middle- Income Countries. Washington: United States Overseas Cooperative Development Council. Retrieved 14 January 2011, from https://www.nreca.coop/wp-content/uploads/2013/07/metrics-3.pdf

Menzani, T., & Zamagni, V. (2010). Cooperative networks in the Italian economy. *Enterprise and Society, 11*(1), 98–127.

Merrien, A. (2014). The socio-economic impact of cooperatives and mutuals. Retrieved February 13, 2015, from https://www.usherbrooke.ca/irecus/fileadmin/sites/irecus/documents/impact_socio-economique_coops_mutuelles/IRECUS-Socio-economic_impact_of_coops_and_mutuals.pdf

Nannyonjo, J. (2013). Enabling agricultural cooperatives through public policy and the state: The case of Uganda from http://www.unrisd.org/80256B42004CCC77/(httpInfoFiles)/FA3911B9091BF39BC1257B7200356723/$file/Justine%20Nannyonjo.pdf

Nilsson, J., Svendsen, G. L. H., & Svendsen, G. T. (2012). Are large and complex agricultural cooperatives losing their social capital? *Agribusiness, 28*(2), 187–204. doi:10.1002/agr.21285.

Novkovic, S. (2008). Defining the co-operative difference. *The Journal of Socio-Economics, 37*, 2168–2177.

Nyiraneze, I. (2007). *Small-scale farming, marketing and organisational support received since 2002 on the Mooi River irrigation scheme in Muden, Central KwaZulu-Natal.* (Master of Agriculture (Food Security)), University of KwaZulu-Natal, Pietermaritzburg.

Porter, P., & Scully, G. (1987). Economic efficiency and cooperatives. *Journal of Law and Economics, 30*, 489–512.

Smith, S. C., & Rothbaum, J. (2013). *Cooperatives in a global economy: Key economic issues, recent trends, and potential for development*. IZA Policy Paper.

Stikkers, K. W. (2011). Dewey, economic democracy, and the Mondragon cooperatives. *European Journal of Pragmatism and American Philosophy, 3*(2), 186–199.

Torgerson, R. E., Reynolds, B. J. & Gray, T. W. (1997). *Evolution of cooperative thought: Theory and purpose*. Paper presented at the Cooperatives: Their Importance in the Future of the Food and Agricultural System, Las Vegas.

United Nations. (1996). Status and role of cooperatives in the light of new economic and social trends. *A/51/267 1996.* Retrieved June 5, 2011, from http://www.un.org/documents/ga/docs/51/plenary/a51-267.htm

University of Wisconsin Center for Cooperatives. (2012). Cooperatives in the U.S. Retrieved March 16, 2014, from http://www.uwcc.wisc.edu/whatisacoop/History/

Von Ravensburg, N. G. (2009). *Enterprise future lies in cooperation—entrepreneur cooperatives in Africa coop AFRICA working paper Dar es Salaam*: International Labour Organisation.

Wanyama, F., Develtere, P., & Pollet, I. (2009). Reinventing the wheel? African cooperatives in a liberalized economic environment. *Annals of Public and Cooperative Economics, 80*(3), 361–392.

Williamson, O. E. (2005). Transaction cost economics. In C. a. S. Memard, M. (Ed.), *Handbook of new institutional economics* (pp. 41–65). Dordrecht: Springer.

Zeuli, K., Freshwater, D., Markley, D., & Barkley, D. (2004). Cooperatives in rural community development: A new framework for analysis. *Journal of the Community Development Society, 35*(2), 17–35.

Chapter 2
Contextualising the Cooperative Movement in Africa

Andrew Emmanuel Okem and Anne Stanton

Abstract This chapter undertakes a review of the evolution of cooperatives in Africa. The review notes that the cooperative form of organisation in Africa has its link to pre-colonial Africa although the modern form of internationally recognised cooperative organisations were introduced by the colonial powers. The chapter also notes that although the rationale that underpins the adoption of cooperatives in Africa is similar to those recognised internationally; this has not been the case in the implementation of cooperatives on the continent. The review provides critical insights into factors that affect cooperatives and note how these can be adopted to enhance the success of cooperatives in contemporary Africa. The chapter concludes by noting that despite challenges faced by cooperatives, they could be instrumental in addressing multiple challenges including poverty, unemployment and hopelessness experienced by many across Africa.

Keywords Africa · Collective activities · Cooperative · Poverty · State control

---

Adapted from chapter two of unpublished doctoral thesis "The Implementation of Cooperative Policy: Perceptions from Cooperatives in the uMgungundlovu District Municipality (Kwazulu-Natal, South Africa)" (Okem 2016).

A.E. Okem (✉)
School of Built Environment and Development Studies, University of KwaZulu-Natal, Durban, South Africa
e-mail: andokem@gmail.com; okem@ukzn.ac.za

A. Stanton
School of Social Sciences, University of KwaZulu-Natal, Pietermaritzburg, South Africa
e-mail: stanton@ukzn.ac.za

A.E. Okem (ed.), *Theoretical and Empirical Studies on Cooperatives*,
SpringerBriefs in Geography, DOI 10.1007/978-3-319-34216-0_2

2.1 Introduction and Chapter Overview

The overriding objective of this chapter is to present an overview of the evolution of the cooperative movement in Africa. The chapter shows that cooperatives in Africa, despite their geographical and historical disparities, share some common fundamental and defining characteristics. This chapter begins by providing a brief overview of the origin of the cooperative movement. This is followed by an examination of cooperatives in Africa in four historical moments: pre-colonial Africa, Colonial Africa, Post-colonial Africa and the era following the implementation of the Structural Adjustment Programmes (SAPs). This historicity provides an understanding of the evolution of cooperatives and how history shapes current practices in relation to the development of cooperatives on the continent. Additionally, lessons could be learnt from the various eras about factors that facilitate the development of cooperatives and those that inhibited their growth. Such lessons could be invaluable in supporting the growth and development of cooperatives in Africa.

2.2 The Origin of the Cooperative Movement

A review of the history of organised cooperatives shows that there have been various attempts at the formation of cooperative societies. Such attempts include the Shore Porters Society established in Aberdeen in 1498 and the Fenwick Weavers Society established in 1761 in Scotland (Mazzarol 2009). However, the Rochdale Society of Equitable Pioneers (henceforth the Rochdale Society), formed in 1844, is often seen as the starting point of present day cooperatives (Mazzarol 2009; Satgar 2011; Towsey 2010). The Rochdale Society emerged during the industrial revolution in England (Mazzarol 2009; Satgar 2007a; University of Wisconsin Center for Cooperatives 2012). The emergence of modern cooperatives at this historical juncture has been construed as a reaction to the harsh socio-economic conditions triggered by the imperatives of the industrial revolution (Ajayi 2012; De Peuter and Dyer-Witheford 2010; Diamantopoulos 2012; Hannan 2014).

During the industrial revolution, skilled artisans, as well as unskilled labourers, lost their jobs as production became increasingly mechanised. In addition, the concentration of capital in the hands of a few industrialists resulted in the pauperisation of many smallholder farmers and artisans who were unable to compete in the industrial age (Jarka et al. 2003; Tchami 2007). These individuals had no organisations to defend them against the exploitation of the capitalist system. These factors, among others, constituted the "macro level factors" which precipitated the "wide-spread proletarianisation in Europe during the 19th and early 20th century" in both rural areas and urban centres (Jarka et al. 2003, p. 242). Against this backdrop, it has been argued that cooperatives emerged as a collective response by the poor towards adapting to and mitigating the imperatives of the industrial revolution (International Cooperative Alliance 2010). By merging their resources, members of cooperatives leveraged scale economics.

From the foregoing, it is apparent that the disadvantaged position of small-scale producers, coupled with the loss of employment by factory workers were key factors that stimulated the formation of cooperatives. According to the Cooperative Development Institute (Cooperative Development Institute 2011), the cooperative model of economic organisation was seen as the only viable means to protect the collective interests of the poor and vulnerable. Similarly, Jarka et al. (2003, p. 242) argue that as a "counter-reaction to capitalistic exploitation", cooperatives brought together disadvantaged peasants who pooled their resources to increase their bargaining power. In the same vein, De Peuter and Dyer-Witheford (2010) note that cooperatives emerged in the late 19th century in opposition to the exploitative capitalist system.

Robert Owen has been credited as an influential thinker whose ideas were at the forefront of advancing the cooperative movement (Ajayi 2012; Satgar 2007a). His thinking was informed by the desire to form a utopian society premised on egalitarianism (Mazzarol 2009). This ideological standpoint facilitated the rapid growth of the cooperative movement in 19th century Europe. The Rochdale Society, which first started out by operating small consumer stores through which members bought basic supplies such as flour and sugar, expanded into other sectors such as housing and finance (Zeuli and Cropp 2004). Since their emergence, cooperatives have played an important role in improving the socio-economic status of people around the world.

2.3 Epochs in the Evolution of Cooperatives in Africa

In Africa, the cooperative movement has evolved over the years. Historically, the evolution of cooperatives on the continent can be placed into four historical periods: pre-colonial Africa, colonial Africa, post-colonial Africa and the era of liberal capitalism. Each era will be presented in the following sub-sections.

2.3.1 Cooperatives in Pre-colonial Africa

The first epoch of cooperatives coincides with pre-colonial Africa during which cooperatives were based on an ideology of self-help. In pre-colonial Africa, the attainment of collective goals was at the forefront of communal activities in recognition of the interdependence of persons in society. The idea of interconnectedness in African society is expressed in Mbiti's maxim "I am because we are. And since we are therefore, I am" (Mbiti 1969, p. 145). This maxim implies that cooperation is an inevitable reality in African societies since each individual sees and realises him/herself through others. Seen from this perspective, cooperation could be construed as a necessary component of African life. Through cooperation, members of pre-colonial African societies shared scarce resources and were able to provide some form of insurance for themselves. This form, although different from cooperatives as per the definition, are underpinned by similar principles and values.

Cooperatives in pre-colonial Africa took many forms including collective management of farm holdings as well as grazing fields. Some elements of pre-colonial forms of cooperatives are evident in Africa today and are manifested in various forms of communal and collective activities. Braverman et al. (1991) have noted some examples of these traditional practices including:

> [...]rotating savings and credit associations (also known as "tontines" or "esusu" in West Africa) that include an element of mutual social assistance in addition to the savings and credit aspect; burial societies, which can be considered as a form of micro-insurance; and mutual work-sharing schemes for large, labour-intensive ventures such as house construction, land clearing or crop harvesting.

Despite the predominance of cooperatives introduced in Africa by the colonial powers, pre-colonial forms continue to persist. For example, a vestige of pre-colonial forms of cooperatives can be found among the Idoma people of central Nigeria in a traditional saving practice known as *otataje* (Ayodele and Arogundade 2014; Iganiga 2008) [a local collective saving society] and collective farming practices called *oluma*. *Oluma* is a practice whereby people organise themselves into groups such as clans or age grades and take turns to work in each other's fields. Through this practice, farmers are able to own and manage large farm holdings beyond what they would be able to manage individually. In addition to working on farms collectively, *oluma* also provides various forms of social support for members in the ebbs and flow of life such as childbirth or loss of a family member. In this way, *oluma* aims to meet both members' economic and social needs. This demonstrates that this traditional practice, although not formally recognised as cooperatives, imbibes values similar to those of formally recognised cooperatives.

2.3.2 Cooperatives in Colonial Africa

The second era in the evolution of cooperatives in Africa coincides with the colonial era. This era was characterised by the introduction of present day cooperatives (Braverman et al. 1991). The driving force for cooperatives during this period was the advancement of the economic interests of colonial powers and cooperative was the exclusive preserve of colonial powers (Nyagah 2012). In colonial Africa, colonial powers tailored cooperatives primarily towards the production of cash crops for their home countries at the expense of subsistence produce (Satgar 2007b; Wanyama 2009). For this reason, the structure and form of support given to cooperatives in colonial Africa encouraged the production of cash crops such as tea, cocoa, coffee and cotton for export.

White owned cooperatives in colonial Africa were heavily subsidised by the then governments (Van Niekerk 1998). The subsidies granted to cooperatives served two main functions. Firstly, they ensured that cooperatives had a monopolistic control over the sector in which they operated. Secondly, they encouraged the settlement of Europeans in the colonies thus strengthening the grip of colonial powers over these colonies (Wanyama 2009). In British colonies, for instance, the British government supported the development of cooperatives among white settlers to generate income for administering the colonies and to provide raw materials for its burgeoning industries. The establishment and expansion of cooperatives in colonial Kenya aimed to benefit white settlers and the Great Britain (Gyllström 1991; Muthuma 2011; Nyagah 2012). The Kenyan scenario demonstrates that the establishment of cooperatives was solely for the benefits of British settlers and the advancement of the economic interest of the British Crown (Nyagah 2012). Effectively, the cooperative movement in colonial Africa were not bottom-up organisations established to meet members' collective interests. In light of the foregoing, Develtere et al. (2008, p. 11) observe that cooperative in colonial Africa

> was not seen as an independent socio-economic movement based on self-managed cooperative enterprises. Cooperatives and the cooperative sector as a whole were treated as instruments for propagating public economic and social policy. In the same vein, members joined cooperatives as matter of public policy rather than the result of voluntary individual motivation. Members belonged to a cooperative either to avoid problems with colonial authorities or to get access to certain services like marketing their produce through the only available channel. They did not regard themselves as the owners of the cooperatives. Consequently, the seeds for a system of cooperatives without co-operators were sown

Although the foregoing paints a general picture of cooperatives in Africa, it is important to note that the colonial approach to cooperatives was not homogeneous. Given the different colonial systems that were operational in Africa, the nature and structure of cooperatives differed across the continent (Develtere et al. 2008). The approach to colonialism adopted by the different colonist influenced the approach to cooperatives. For instance, the social economic model was predominant in French colonial territories while the British adopted a unified approach to cooperatives. In Portuguese colonies, the producer model was implemented in contrast to the social movement model in Belgian colonies. Countries such as Egypt, South Africa, and Ethiopia adopted an indigenous model in the implementation of cooperatives (Develtere et al. 2008).

2.3.3 Cooperatives in Post-colonial Africa

The end of colonialism brought about different social and political imperatives, which led to the third phase in the evolution of cooperatives in Africa. During this period, leaders of newly independent African countries saw cooperatives as a means of improving social cohesion and fast-tracking economic development in their respective countries (Getnet and Anullo 2012). From this ideological standpoint, cooperatives in post-colonial Africa were seen as extensions of the state (Satgar 2007a). In this respect, the notion of cooperatives promoted by first generation African leaders was similar to those of the colonial era. First generation African leaders made no attempts at changing the organisational forms of cooperatives.

During the third phase of the evolution of cooperatives in Africa, states were actively involved in supporting the development of cooperatives (Satgar 2007b). Seen as extensions of the state, cooperatives were not perceived as independent and self-organising enterprises aimed at improving members' collective interests. Rather, they were subjected to the control of state institutions. In Kenya, for instance, The Co-operative Societies' Act (CAP 490 of 1966) placed the cooperative movement firmly under government control. Massive financial injection into the country's cooperative sector and the establishment of a Cooperative Department to facilitate and manage the development of cooperatives were all directed towards government control of cooperatives. Because of the emphasis on agricultural cooperatives during this period, government policy also translated into the rolling out of extensive agricultural extension services.

A common approach to cooperatives in post-colonial Africa was that they were conceived as a paradigm for the advancement of states' policies—particularly in the area of local economic development (Hartley and Johnson 2014). For instance, the socialist policy of Julius Nyerere's Regime in Tanzania effectively placed cooperatives under state control. According to Wanyama (2012), Nyerere saw cooperatives as an ideal tool for the implementation of his *Ujamaa* policy. Consequently, government support of cooperatives was essentially dependent on cooperatives' ability to implement government's theory of change[1] (Satgar and Williams 2008). In doing this, government eroded the autonomy of cooperatives. In addition, the actions of cooperatives were no longer geared towards achieving members' interests; rather they aimed primarily to attain governments' vision for cooperatives.

In Ethiopia, which did not experience institutionalised colonialism, the pattern of the cooperative movement during this period was similar to that of most African countries. The military rule from 1975 to 1991 provided extensive institutional support which led to the proliferation of cooperatives. These cooperatives were not underpinned by the passion to address members' collective needs. Rather, they were collective responses to government financial incentives. Consequently, the end of the military rule and associated finance support of the sector resulted in the collapse of most cooperatives which relied on government support for their existence (Japan Association for International Collaboration of Agriculture and Forestry 2008). The collapse of most cooperatives at the end of military rule demonstrated that most cooperatives established during the military regime were not underpinned by the principles and values of the cooperative movement.

[1]In the development field, a theory of change refers to the methodology used to bring about change. The theory, which often emphasises the value of participation, links the causal pathways that must be followed in order to achieve certain predetermined outcomes. According to Vogel (2012: 9), theory of change is underpinned by "evaluation and informed social action".

State control of cooperatives had detrimental effects on the growth of cooperatives across the continent. Among other things, the state-led approach to the development of cooperatives was characterised by a series of abuses of the cooperative model. Such abuses include undermining the autonomy of cooperatives, creating a strong patronage system that made cooperatives dependent on the state, absence of democratic member control as well as overbearing government bureaucracies (Satgar and Williams 2008). The foregoing is contrary to the internationally recognised principles and values of the cooperative movement (International Cooperative Alliance 1995).

In post-colonial Africa, cooperatives did not subscribe to principles such as member economic participation or concern for community. Seen as extensions of the state, members of cooperatives had little incentive to work towards the promotion and sustainability of cooperatives. Cooperatives in this period relied on government for funds, governance, marketing, and training. Extensive reliance on state resources meant that the failure of state institutions often had a direct bearing on the failure of cooperatives since they were intrinsically linked to state bureaucracies most of which were plagued by inefficiencies and maladministration (Wanyama et al. 2009). Herein lies the paradox of cooperatives in Africa. While cooperatives espouse the value of independence, they often rely on the state for their continued existence. This reliance entails the creation of state support institutions for cooperatives. In this scenario, the distinction between what constitutes state control and state support is often blurred.

2.3.4 Cooperatives in Liberal Economic System in Africa

The link between the failure of state institutions and the failure of cooperatives has been identified as one of the motivating factors for disengaging cooperatives from the state in the late 1980s and early 1990s (Wanyama et al. 2009). Thinking about making cooperatives independent of the state was informed by the conviction that they are more successful if they operate independently of government institutions (Wanyama et al. 2009; Zeuli and Cropp 2004). Additionally, delinking cooperatives from the state reflects a commitment to institutionalising the principles and values of the cooperative movement.

The push for the independence of cooperatives in Africa re-emphasised the values of cooperation in the pre-colonial era when it was underpinned by the value of self-help in the absence of formal state support. However, this value gave way to state control during the colonial era. From pre-colonial Africa to this period, cooperatives have progressed from self-help to state control, and to state reliance. Despite these changes, the objective remained that of social and economic development even though the nature thereof was different at different times (subsistence in the pre-colonial era, extractive in colonial times for the benefit of colonisers, and state-centred after independence for the benefit of state leaders).

The call for independent, people-centred cooperatives in the late 1980s and early 1990s was about enabling cooperatives to realise the values of their existence. According to Berolsky (2000) and William (2003), such thinking fits into the broader policy intervention of the Bretton Woods Institutions in the form of the Structural Adjustment Programmes (SAPs) of the 1990s. The SAPS advocated a capable state in place of a large and bloated state. In addition, it encouraged the implementation of a *laissez-faire* capitalist system in which the state has limited participation in the economic sphere. The implementation of neo-liberal economic policy led to the rapid privatisation of state-owned enterprises, deregulation of various sectors of the economy and minimal state intervention in the economic sphere (Berolsky 2000). Liberalisation, occasioned by the implementation of the SAPs, resulted in the removal of government subsidies. In addition, it led to the end of price control as well as the drastic restructuring of public service to reduce bloated bureaucracy. Drastically reducing the size of bureaucracies correspondingly reduced government support structures for cooperatives.

The socio-economic imperatives of the SAPs were a stimulus for the emergence of the fourth era of the evolution of cooperatives in Africa. What was apparent during this era was less state interference in the activities of cooperatives. This period was characterised by the formulation and implementation of policies that saw cooperatives as a movement independent of government institutions (Wanyama et al. 2009). In addition, the role of members in advancing the success of cooperatives was emphasised over that of the state. Rather than being seen as extensions of the state, cooperatives are seen as "democratically and professionally managed, self-controlled and self-reliant" organisations (Wanyama et al. 2009, p. VI).

Although making cooperatives independent of the state was seen as a positive initiative as far as the development of the cooperative movement was concerned, cooperatives in many African countries performed poorly during this period (Ajayi 2012; Hartley and Johnson 2014). For instance, a study by Wanyama et al. (2009) found that economic liberalisation had a considerable negative impact on cooperatives because cooperatives, for many years, relied heavily on state institutions and were ill-prepared for the new economic climate brought about by the SAPs. An outcome of these changes was the collapse of cooperatives that could not function in the absence of privileges and support they previously enjoyed. In East Africa for example, liberalisation of the coffee market resulted in the entrance of multinational corporations into the domestic market. These corporations significantly weakened the market share of cooperatives in the coffee sector (Ponte 2002). Despite this, Wanyama et al. (2009) argue that the economic reality brought about by the SAPs was a positive development for cooperatives on the continent because it resulted in the elimination of ineffective cooperatives that relied solely on state subsidies for survival. It has also led to the emergence of new institutional cooperative models that provide market support for local cooperatives in countries such as Ethiopia & Kenya (Kodama 2007).

Apart from attempting to eliminate 'wasteful government expenditures' (one of the underlying philosophies of the SAPs), liberalisation also ensured that cooperatives were pressured into finding innovative ways to adapt to the new economic climate. In addition, it created a condition in which the formation of cooperatives was no longer hinged on the prospect of accessing government support in the form of subsidies and grants.[2] Rather, cooperatives were seen as organisations for advancing members' social and economic interests (Wanyama et al. 2009). Furthermore, Satgar and Williams (2008) argue that the success of cooperatives in Africa, following the introduction of the neo-liberal policy, was underpinned by peoples' passion for the cooperative movement. Consequently, this era played an invaluable role in ensuring that cooperatives in Africa embody the principles and values of the cooperative movement. Government approach to cooperatives, following the implementation of SAPs, continue to underline government relation with cooperatives across Africa. However, the extent to which cooperatives operate independently of government interference differs across countries.

2.4 Lessons for the Review of the History of Cooperatives in Africa

The historicity of cooperatives as outlined above provides an understanding of the evolution of cooperatives in Africa and how history shapes current practices in relation to the development of cooperatives on the continent. Additionally, the review provides lessons about factors that facilitate the development of cooperatives and those that inhibit their growth in a given era. An important lesson derived from this review is that cooperatives thrive when they operate as independent institutions focused on improving members' socio-economic conditions. In addition, the review showed that over-reliance on the state could result in a weak cooperative sector since their performance is linked with that of supporting government institution(s). Thus, it is argued that although government support can facilitate the growth of cooperatives (often seen in terms of the number of registered cooperatives), government support should be limited to the creation of an enabling environment that will facilitate their growth. In doing this, cooperatives are positioned to contribute to poverty alleviation, employment creation, economic development, and people's empowerment.

2.5 Conclusion

The overriding objective of this chapter was to present an overview of the evolution of the cooperative movement in Africa. The chapter began by examining the emergence of cooperatives. The chapter noted that although modern cooperatives

---

[2]In countries such as South Africa and Nigeria, government grants continues to be the main source of revenue for cooperatives despite the prevalent view that this approach is detrimental to the success of cooperatives.

were introduced during colonialism, cooperatives in Africa predate the colonial era. Pre-colonial forms of cooperative can still be found today despite the predominance of those introduced in the colonial era. From the historical overview of the evolution of cooperatives in Africa, the chapter established that cooperatives are more successful in contexts where they are bottom-up organisations that operate independently of government interference. This is evident in the failure of cooperatives in post-colonial African when cooperatives were adopted as tools for the implementation of the socialist ideals of many African leaders. Reflecting on the negative impact of this approach during this period, Restakis (2010, 53) notes that

> …co-operatives became the instrument of choice to implement state policies for production and economic development. Voluntary cooperation was replaced by mandated co-operation. And so it came about that centralized socialism became far more damaging to the integrity of the co-op idea and the realization of its potential than capitalism itself. To this day co-operatives in many of these countries signify little more in the minds of the populace than instruments of state coercion. It is a tragedy of economic and human misuse whose negative effects are still being felt.

For this reason, the role of government should pertain only to the creation of conditions for the emergence and growth of cooperatives.

References

Ajayi, J. R. (2012). *Strategies for sustainable housing co-operatives in South Africa.* (Doctor of Philosophy (Construction Management)), Nelson Mandela Metropolitan University, Port Elizabeth.

Ayodele, A. E., & Arogundade, K. (2014). The impact of microfinance on economic growth in Nigeria. *Journal of Emerging Trends in Economics and Management Sciences, 5*(5), 397–405.

Berolsky, N. (2000). *An evaluation of IMF structural adjustment programmes: Lessons for South Africa.* (Master Of Social Science), Rhodes University in Fulfilment. Retrieved from http://eprints.ru.ac.za/2172/1/BEROLSKY-MSc-TR00-06.pdf.

Braverman, A., Guasch, L., Huppi, M., & Pohlmeier, L. (1991). *Promoting rural cooperatives in developing countries: The case of Sub-Saharan Africa World Bank discussion paper.* Washington: The World Bank.

Cooperative Development Institute. (2011). History of cooperatives: The early origins of cooperatives. Retrieved 16 March 2014, from http://www.cdi.coop/historyofcoops.html.

De Peuter, G., & Dyer-Witheford, N. (2010). Commons and cooperatives. *Affinities: A Journal of Radical Theory, Culture, and Action, 4*(1).

Develtere, P., Pollet, I., & Wanyama, F. (Eds.). (2008). *Cooperating out of poverty the renaissance of the African cooperative movement.* Geneva: International Labour Office.

Diamantopoulos, M. (2012). Breaking out of co-operation's 'iron cage': From movement degeneration to building a developmental movement. *Annals of Public and Cooperative Economics, 83*(2), 199–214. doi:10.1111/j.1467-8292.2012.00461.x.

Getnet, K., & Anullo, T. (2012). Agricultural cooperatives and rural livelihoods: Evidence from Ethiopia. *Annals of Public and Cooperative Economics, 83*(2), 181–198. doi:10.1111/j.1467-8292.2012.00460.x.

Gyllström, B. (1991). *State administered rural change: Agricultural cooperatives in rural Kenya.* London: Routledge.

Hannan, R. (2014). The institution of co-operation: A pathway from governance to spillover and poverty reduction. *Journal of Co-operative Organization and Management, 2*(1), 34–42. doi:10.1016/j.jcom.2014.03.003.

Hartley, S., & Johnson, H. (2014). Learning to co-operate: Youth engagement with the co-operative revival in Africa. *European Journal of Development Research, 26*(1), 55–70.

Iganiga, B. (2008). Much ado about nothing: The case of the Nigerian microfinance policy measures, institutions and operations. *Journal of Social Sciences, 17*(2), 89–101.

International Cooperative Alliance. (1995, 26 May 2007). Statement on the Co-operative Identity. 15 March, from http://www.ica.coop/coop/principles.html.

International Cooperative Alliance. (2010). *Global 300 report 2010 international co-operative alliance: The world's major co-operatives and mutual businesses*. Geneva: International Cooperative Alliance.

Japan Association for International Collaboration of Agriculture and Forestry. (2008). *One village one product movement in Africa issues and feasibility in Ethiopia*. Akasaka: Japan Association for International Collaboration of Agriculture and Forestry.

Jarka, C., Gunnar, L. H. S., & Gert, T. S. (2003). Building and destroying social capital: The case of cooperative movements in Denmark and Poland. *Agriculture and Human Values, 20*, 241–252.

Kodama, Y. (2007). New role of cooperatives in Ethiopia: the case of Ethiopian coffee farmers cooperatives. *African Study Monographs, Supplement, 35*, 87–108.

Mazzarol, T. (2009). Cooperative entreprise: A discussion paper and literature review. Retrieved 14 January 2013, from http://www.cemi.com.au/sites/all/publications/0901%20COOPS%20WA%20LIT%20REVIEW.pdf.

Mbiti, J. (1969). *African religion and philosophy* (2nd ed.). New Hampshire: Heinemann.

Muthuma, E. W. (2011). *Economic cooperation in Kenyan credit cooperatives: Exploring the role of social capital and institutions.* (Doctor of Philosophy), University of the Witwatersrand, Johannesburg.

Nyagah, W. N. (2012). *Co-operatives as potential channel for enhancing financial inclusion.* Abuja, Nigeria: Paper presented at the Enhancing Financial Innovation and Access.

Ponte, S. (2002). Brewing a bitter cup? Deregulation, quality and the re-organization of coffee marketing in East Africa. *Journal of Agrarian Change, 2*(2), 248–272. doi:10.1111/1471-0366.00033.

Satgar, V. (2007a). Cooperative development and labour solidarity: A neo-gramscian perspective on the global struggle against neoliberalization. *Labour, Capital and Society, 40*(1–2), 56–79.

Satgar, V. (2007b). *Neoliberal restructuring and primitive accumulation labour and the global struggle for cooperative alternatives*. Paper presented at the Global Labour University (Thrust: Labour, Growth and Development). Retrieved 16 March 2014, from http://www.global-labour-university.org/fileadmin/Papers_Wits_conference_2007/A2/satgar_paper.pdf.

Satgar, V. (2011). Challenging the globalized agro-food complex: Farming cooperatives and the emerging solidarity economy in South Africa. *WorkingUSA, 14*(2), 177–190.

Satgar, V., & Williams, M. (2008). The Passion of the People: Successful Cooperative Experiences in Africa: Cooperative Alternative Policy Center.

Tchami, G. (2007). Handbook on cooperatives for use by Workers' Organizations *Handbook on cooperatives for use of Workers' Organizations* (Vol. Cooperative Programme). Geneva: International Labour Office.

Towsey, M. (2010). The biopsychology of cooperation. *Understanding Proutt–Essays on Sustainability and Transformation, 1*.

University of Wisconsin Center for Cooperatives. (2012). Cooperatives in the U.S. Retrieved 16 March 2014, from http://www.uwcc.wisc.edu/whatisacoop/History/.

Van Niekerk, J. A. S. (1998). *Cooperative theory and practice*. South African Agricultural Union (now known as AgriSA): Pretoria.

Wanyama, F. (2009). *Surviving liberalization: The cooperative movement in Kenya.* Geneva: International Labour Organization.

Wanyama, F. (2012). *The social and solidarity economy in African development: The role of cooperatives*. Paper presented at the Cooperative Symposium on "Perspectives for Cooperatives in Southern Africa", Lusaka, Zambia.

Wanyama, F., Develtere, P., & Pollet, I. (2009). Reinventing the wheel? African cooperatives in a liberalized economic environment. *Annals of Public and Cooperative Economics, 80*(3), 361–392.

William, E. (2003). *IMF and World Bank structural adjustment programs and poverty*. Chicago: University of Chicago Press.

Zeuli, K. A., & Cropp, R. (2004). Cooperatives: Principles and practices in the 21st century. Retrieved 19 January 2013, from http://learningstore.uwex.edu/assets/pdfs/A1457.PDF.

Chapter 3
A Review of Possible Emerging Theoretical Debates and New Interdisciplinary Perspectives on the Cooperative Movement in Africa

Gordon Fakude

Abstract The remarkable resilience displayed by the cooperative sector in international experience is also manifested in the African continent where cooperatives can be said to have traversed a particularly difficult trajectory. Despite the commonly acknowledged fact that most post-colonial African governments attempted to use cooperatives as tools of development in their quest to attain nation-building goals and did not allow them to become fully autonomous member-owned and controlled businesses, there is evidence of a resurgence of cooperatives in the continent. Several studies have examined conditions under which cooperatives thrive or fail. These studies have brought to the fore interesting insights on a number of critical factors; key amongst them is the role of social capital in the creation of a facilitative environment conducive to the growth of cooperatives. This chapter will interrogate this literature with a view to understanding not only the key theoretical debates but also the interdisciplinary perspectives that are shaping the intellectual discourse accompanying the resurgence of the cooperative movement in Africa.

Keywords Cooperative · Interdisciplinary perspectives resilience · Resurgence · Social capital

3.1 Introduction and Chapter Overview

This chapter seeks to review recent studies and the emergence of interdisciplinary perspectives on the cooperative movement in Africa. The chapter is based on the review of recent studies on the cooperative sector in Africa in the

G. Fakude (✉)
Economic History and Development Studies, School of Social Sciences,
University of KwaZulu-Natal, Durban, South Africa
e-mail: fakude@ukzn.ac.za

A.E. Okem (ed.), *Theoretical and Empirical Studies on Cooperatives*,
SpringerBriefs in Geography, DOI 10.1007/978-3-319-34216-0_3

post-liberalisation era.[1] This review draws from several country studies in the past decade focusing on various dimensions of the sector. Its point of departure in the first section unpacks the historicity of the cooperative movement by briefly tracing the evolution of cooperatives in Africa since their establishment in the early decades of the 20th century. Perspectives on this history are reviewed to throw light on current debates on how this history shapes the current approaches to the development of cooperatives on the continent. The next section extends this review to include an exploration of selected topical themes that are emerging in interdisciplinary perspectives on the cooperative movement in Africa and focuses especially on the perceived potential of cooperatives to contribute to poverty reduction on the continent.

3.2 A Resurging African Cooperative Sector

Recent studies confirm that a remarkable resurgence of cooperatives is being experienced across the African continent with members of these organisations taking control of their fate as they chart new development trajectories (Wanyama 2008; Theron 2008; Mendonça 2008; Nyamwasa 2008; Satgar and Williams 2008; Pollet 2009). In this resurgence, new relationships are being forged and old ones are being revived or abandoned under an agenda that is no longer solely determined by the state and/or donors (as it used to happen in the past), but an agenda that is simultaneously seeking to safeguard members' interests (Develtere et al. 2008; Satgar and Williams 2008).

Interpretations of this resurgence, however, differ. On one hand, there is a view that, at last, the cooperative sector in Africa is experiencing a fresh start where, under strong leadership and good management, they are "beginning to prosper as autonomous businesses they should always have been" (Birchall 2003: 12). A common perception in this view is that of the irrelevance of past history, particularly the history of cooperatives during the post-colonial era when cooperatives existed under state tutelage. This history is usually dismissed as fraudulent by authors such as Birchall (2004: 16), who, for instance, following Munkner (1989) before him,[2] concludes that "the history of cooperatives in developing countries between 1950s and 1980s can legitimately be rejected as 'their story' and 'not ours' (i.e. not the story of a component of the world's cooperative movement)". A different view is, however, emerging from recent studies which have brought to

[1]The post-liberalization era (in African economies) is generally assumed, in literature, to commence soon after the epochal events such as the fall of the Berlin Wall and occurred in parallel with the democratization wave in the 1990s that saw the end of one-party rule in most African states.

[2]Munkner (1989) referred to the cooperative sector during colonial and post-colonial eras in Africa as a system of cooperatives without co-operators as members of cooperatives did not regard themselves as owners of the cooperatives.

light evidence that cooperatives in Africa are not making a fresh start really but what is happening is that with the resurgence of interest in cooperative enterprises among grassroots groups, government agencies, and even donors, old cooperatives are being revived while new ones are being created at a rapid pace (see Develtere et al. 2008; Satgar and Williams 2008; Pollet 2009; Allen and Maghimbi 2009). In this vein, Develtere et al. (2008: 1) argue that all parties involved in this resurgence or renaissance of cooperatives in Africa "are operating in an environment that is to a large extent culturally and institutionally influenced by past developments".

The different interpretations do not signify a conflicting understanding of history but are rather a reflection of the numerous perspectives particularly of the recent history of the post-colonial state interventionism in African economies. The mainstream view, from which the former view derives, has always emphasised that the blame for the poor performance of African economies rests squarely on the misguided policies of state intervention in economic development pursued by African states. Abundant reports of cooperatives' failures reinforced this perception of the failure of state interventionism. This led to the castigation of the model of collective enterprise ownership to such an extent that the history of cooperatives in the post-colonial era of state interventionism began to be equated with some fraudulent activities of the post-colonial state (as shown in the above comments of some protagonists of this view). The World Bank studies of the 1990s came to a similar conclusion that many cooperatives in Africa had become fraudulent derivatives of the international cooperative tradition as they were functioning as semi-public bureaucratic enterprises and not genuine, voluntary and private businesses (Hussi et al. 1993; Porvali 1993). Nonetheless, the World Bank studies underlined that cooperatives remained the preferred form of organisation and the only organisation with which rural people were familiar.[3]

A different view to the dominant and mainstream view expresses sentiments that are sympathetic to the perspective that has sought to locate the poor performance of African economies in the broad context of existing skewed relationships in the global economy where the *dirigiste* mode of operation of the post-colonial state, especially with regard to cooperatives, can be understood as being a legacy from the coercive colonial regimes. Develtere et al. (2008) highlight the disconnect that existed between the motivation (of the colonial state) to form cooperatives and the interests of the African people that led to the pattern of cooperative development directed from above by the state which was inherited and accentuated by the post-colonial state. These researchers point out that the persistence of this pattern was largely because cooperatives, in both the colonial and post-colonial eras, were treated by authorities as instruments for propagating economic and social policies and were not seen as independent self-managed enterprises that they were in metropolitan countries. In response to this subtly coercive environment, people

---

[3]The World Bank studies concluded that there was a potential role that cooperatives (in Africa) could play in the development process, provided they were disentangled from the state and restructured along business principles in line with the market economy (Hussi et al. 1993; Porvali 1993).

joined cooperatives to avoid conflicts with authorities and, under post-colonial regimes, to get access to certain services like marketing their produce through the only available channel (Develtere et al. 2008: 11). The authors observe that cooperative members did not regard themselves as owners of cooperatives and consequently, the seeds for a system of cooperatives without cooperators that Munkner (1989) refers to, were sown.

It is worth noting that notwithstanding these differences in perspectives, there is no dispute on the cooperatives' historicity and, in fact, there is convergence in recent research. It can be argued that although cooperatives in Africa were set up from foreign models to further exterior motives, they have evolved over time, warts and all in their history, to become "one of the main forms of popular economic and social organization for the alleviation of poverty" (Satgar and Williams 2008: 5). The research on cooperatives so far reveals that cooperatives have not withered away with the liberalisation measures. They continue to play an important role in many African economies. Studies confirm an average penetration rate of seven per cent (i.e. meaning that on average about seven percent of the African population belongs to a cooperative) although there are countries (e.g. Egypt, Senegal, Ghana, Kenya, and Rwanda) that report a higher penetration rate of over ten percent (Develtere and Pollet 2008: 46; Wanyama et al. 2009: 11). Satgar and Williams (2008: 5) attribute the motivation for this renaissance to a 'passion of the people', as they attempt to explain the resilience that has been displayed by the cooperative sector in Africa despite strong state control of the post-colonial era and the "market appropriation of cooperatives by neoliberal restructuring" attendant on structural adjustment's liberalisation of developing countries' economies. It would be useful if this study explained what constituted the basis for this 'passion of the people' which obviously could be linked to the solidarity traditions that have always existed in African people's cultures.

3.3 Path Dependence and Cooperatives' Development Trajectory

In their research, Develtere et al. (2008: 1)[4] found that the fourth generation cooperatives emerging in post-liberalisation Africa carry a heavy legacy that conditions their development path. Their study confirmed that in spite of the advent of a more competitive market environment that followed the liberalisation of economies in conjunction with a more democratic political opportunity structure that made it possible to re-envisage cooperatives as private sector agents primarily driven by the

[4]The author of this review concedes that the perception that this is an analysis of the Develtere et al. (2008)'s study (rather than a review that is meant to bring new and original information), may be unavoidable given the iterative references to that study. Repeated reference to Develtere et al. (2008) study is largely due to the fact that in the past decade this is the only major study with a continent-wide coverage that has been conducted.

interests of members, the resurgence and the new dynamism currently experienced does not mean that there is a fresh start in the cooperative sector. According to these researchers, the colonially bequeathed models, namely, the British-inspired unified cooperative model; the social economy model (that was historically found in French-ruled territories); the social movement model (used predominantly by the Belgians); the producers' model (introduced by the Portuguese in their African colonies); and an indigenous model (South Africa, Egypt and Ethiopia), continue to survive in cooperative policy, cooperative organisation, cooperative networking and cooperative culture. Thus, a rich and diversified cooperative scenery is found in Africa today.

This diversified cooperative scenery is explained by the researchers in this study through the concept that social scientists call "path dependence". Path dependence is defined as a situation where a route to be taken is determined by antecedent conditions and past choices or decisions (Develtere et al. 2008: 2). With regard to cooperatives, Develtere et al. (2008) maintain that path dependence was first manifested when post-colonial states decided not to alter the ties of the cooperative development strategy to the export crop strategy which colonial agencies had put in place. They opted for the reinforcement of this cooperative-export nexus by making cooperatives sub-contractors of the marketing boards. These researchers argue that in the post-liberalisation phase, path-dependence is still manifested in the cooperative structures and networks that are being set up. One observation is that, whilst in some countries, particularly in the Anglophone world, there are still attempts to unite and streamline the sector as envisaged in the unified cooperative model, in many other Anglophone countries, new forms of cooperation are now accepted and even legally recognised. This confirms that the social economy model is making inroads and is being adopted in many countries. At the same time, cooperatives are venturing out to new areas such as healthcare, social services, environmental protection and manufacturing as many communities respond to poverty and increasing unemployment by forming cooperatives.

Country case studies on which the research by Delvetere et al (2008) is based also confirm inconsistencies in policy formulation which may be partly explained by path dependence. Whilst most countries have moved from heavy involvement in the cooperative sector towards a situation where this is less so, cooperatives policy scenario is still very erratic as policy lines and programmes are developed in an inconsistent way (Develtere and Pollet 2008: 68). Researchers thus highlight that the degree of distance that the government and administration now have from the cooperative sector differs from country to country. These differences range at one end of the spectrum from the Cape Verdian case where government has completely abandoned the sector and left it to its own regulation which has left the sector in disarray; to government retreat in Uganda where some cooperatives collapsed as a result but at the same time new cooperatives have emerged in non-traditional areas (vanilla and oil-seed); to the Ghanaian case where government's attempts to re-orient cooperatives to work in a liberalised environment are frustrated by the lack of clear government policy; to the more facilitative stance that has been adopted by the Kenyan government following the initial fall into decay of some cooperatives when government released its grip on the cooperative sector. At another end, some governments have taken a more pro-active attitude towards the cooperative sector. This seems to be the case in countries such as South Africa, Rwanda and Ethiopia where there were profound political changes in the 1990s. The pro-active attitude in these countries is reflected in the mild intervention in South Africa, for instance, where the government has established a Cooperative Development Unit and adopted a Cooperatives Development Policy and intends to use cooperatives to bridge the divide between formal and informal sectors of the economy. In Rwanda, on the other hand, the state is encouraging the formation of trade federations with both cooperatives and non-cooperatives (Develtere and Pollet 2008).

3.4 Cooperatives in Africa's Economic Growth and Poverty Reduction

Birchall (2003) contrasts two commonly held views on the potential of cooperatives to reduce poverty. One view which was commonly held by policy makers and some development practitioners was that cooperatives have an automatic tendency to benefit the poor. This was a dominant view held by both the post-colonial states and donors especially in the 1960s when large amounts of aid were channeled to state-sponsored cooperatives by donors. A key assumption upon which this practice was based was that cooperatives were open to new members and did not require people to invest large amounts of capital in setting them up. Thus, poor people could be encouraged to form cooperatives that would enable them to take part in economic activities which could lead to the possibility of earning incomes to improve their lives. In contrast, another view took a minimalist approach that cooperatives were people-centred businesses which were mainly concerned with benefitting their members and did not have any inbuilt obligation to benefit the poor. Protagonists of the latter view held that cooperatives were not designed as tools for poverty reduction, but were a means by which groups of people could gain economic advantages that individually they could not achieve. Between these two positions, Birchall (2003) proposes a balanced view which cautions that a poverty-reduction potential is realised only when cooperative values and principles are respected and that cooperatives can contribute more effectively to poverty reduction than other forms of economic organisation.

This debate continues to resonate in Africa where, as researchers have already pointed out, cooperatives have become one of the main forms of popular economic and social organisation for the alleviation of poverty. According to Develtere et al. (2008: xv), the resurgence of cooperatives in Africa is witnessing a phenomenon where the poor are "cooperating out of poverty". Liberalisation of economies in Africa has afforded space for cooperatives to re-examine their organisational formations and they are doing so primarily with a view to safeguard their interests rather than the interests of the state as was the case in the past. This is what is interpreted by some as the fresh start that is taking place in the cooperative sector in Africa.

In many countries in Africa, cooperatives play a crucial role in the economic and social transformation and they have been found to constitute a parallel sector in the economy (Satgar and Williams 2008: 13). Hence, Satgar and Williams (2008: 13) have identified the cooperative movement as "one of the most organized social forces on the African continent". The ILO data (alluded to earlier) suggests that at least 7 percent of African citizens belong to cooperatives whilst this proportion rises to about 10 percent in countries such as Mauritius, Egypt, Ghana, Kenya, and Senegal. This affirms that cooperatives are an organised social force in Africa.

3.5 Cooperatives' Social Capital and Poverty Alleviation

With regard to the creation and sustenance of social capital,[5] there may be a semblance of truth in the assertion that African cooperatives in this regard are making a fresh start, because for the first time in the history of cooperatives following the economic liberalisation era, people are allowed to network in their interests (as they should always have done since solidarity systems always existed in all countries in spite of the fact that these systems were deliberately ignored when cooperatives were first established). Expectations are that, since the networks are voluntarily created, this will likely lead to the establishment and strengthening of viable and sustainable enterprises and institutions which will help form a sound base for a sustained economic development process. Social capital which is fostered by the networking activities is considered to be the necessary lubricant in development. Historical experience teaches that cooperatives are among the leading economic and social institutions that are suitable for its generation.[6]

[5]There is abundant theoretical literature on the value of social capital in the development process (Aguilar and Sen 2009; Bourdieu 1986; Coleman 1988; Putnam 1995; Portes 1998; Woolcock and Narayan 2000; Grootaert 2001). In this literature there is a growing consensus that social capital stands for "the ability of social actors to secure benefits by virtue of membership in social networks or other social structures" (Portes 1998: 6). This corresponds with the seminal definition of social capital by Putnam (1995) where he defined it as "features of social organization, such as networks, norms, and trust that enable participants to act together more effectively to pursue shared objectives" (pp. 664–665).

[6]The principles that undergird the cooperative structure, i.e. open and voluntary membership, democratic member control, autonomy and independence, member economic participation, member training and education, cooperation among cooperatives, and concern for community, and the values of self-help, self-responsibility, democracy, equality, equity and solidarity upon which cooperatives are based, embody a conducive configuration for the creation and sustenance of social capital. In this vein, a number of studies have highlighted the case that cooperatives are well placed to help poor people strengthen their participation and inclusion in economic activities at the local level (Woolcock and Narayan 2000; Birchall 2003, 2004; Majee and Hoyt 2009).

In the meantime, as shown above, researchers have confirmed that social capital is a critical asset in the socio-economic well-being of people and that it is even "relatively more important in the portfolio of assets of poor households" (as) "it gives them a relatively higher return" (Grootaert 2001: 22). Majee and Hoyt's (2010: 428) case study on worker-owned cooperatives in the US confirmed the cooperatives' potential to build trust, increase participation and networking among members in business and social activities. They thus contend that these organisations are "well-suited to create a socio-economic context beneficial to both members and community". These researchers also pointed out areas that need further research, e.g. how factors such as location, demographic composition, size, years in operation, etc. influence social capital creation. These area may be of interest especially to policy makers in the African continent as they grapple with what needs to be done to create a conducive environment for cooperatives.

Whilst states and donors may have realised the potential of cooperative structures, a key lesson from the sad history of excessive interference by the state in cooperative affairs is that such interference stifled and suffocated people's initiative and thus hampered the creation of networks necessary for the development process. Unfortunately, that lesson may not be so clear to some governments due to pressures in the political system that are pushing for more intervention by the state in order to achieve certain political objectives. A case in point here is that of post-apartheid South Africa where emerging cooperatives have been identified as the key vehicles or instruments to achieve black economic empowerment.[7] This may not be a bad goal to aim for given the extent of inequalities in the country's socio-economic life and the development imperative of reintegrating the vast majority of black people in the economy to open up opportunities (which they were denied by law in the past). Such an approach will enable them to become active participants in the economy and not just the source of cheap labour as used to be the case under successive white regimes since the advent of colonialism in this part of Africa. The strategy , however, should be pragmatically managed because a key lesson from past experience is that using cooperatives in order to achieve national development objectives is likely to backfire when the values and principles of cooperatives are not respected.

A related lesson observed by Develtere et al. in their 2008 study is that "cooperative ventures best thrive when the norms and values they rely on are congruent with those inherent in pre-existing or parallel systems (even if institutionally they are not congruent or linked)". Heeding such lessons from past experience and incorporating them in the design of support programmes may help avoid pitfalls such as those that have been uncovered by recent research on the use of state's procurement

[7]In this vein, Satgar (2007) observed that "whilst the Broad-Based Black Economic Empowerment (BBBEE) Act of 2003 explicitly provides for cooperatives to be part of socio-economic strategies to empower black people, implementation especially with regard to cooperatives has often been plagued by numerous contradictions: e.g. the get-rich-quickly logic of the BEE discourse in general does not assist with institutionalizing cooperatives in a sustainable way on the ground" (p. 5).

to foster the growth and expansion of the emerging cooperative sector in KwaZulu-Natal province of South Africa. Research has found excesses that led to gross inefficiencies and a waste of resources as the state in KwaZulu-Natal province, for instance, replaced SMMEs with women cooperatives in its school feeding schemes which resulted in the inefficient provision of the service (Ballard and Beesley 2013). In a paper on the state of the cooperative sector in South Africa, Satgar (2007: 14) cautioned that "while the KZN government's approach to ensure state support to (emerging) cooperatives is well intentioned, it was not enabling nor developing cooperatives in accordance with member needs and capacities but was instead instrumentalizing cooperatives according to the state's objectives". A key lesson that has been repeatedly emphasised in literature with regard to the contribution of cooperatives to poverty alleviation is that while the potential of cooperatives to reduce poverty is apparent and that it may be a relatively less costly strategy to engage them, it is important to note that such potential may not be realised (as Africa's post-colonial experience has shown) because those concerned with poverty reduction either "lose sight of the needs" of cooperatives' members or "see cooperatives as tools rather than as autonomous organizations" (Birchall 2003: 4).

In all, the critical role of social capital in the growth and expansion of the cooperative sector has been highlighted by research as a key success factor in the resurgence of cooperatives in Africa. It also revolves around cooperatives' ability to network among themselves and also to establish viable partnerships and create alliances even further afield. This has been witnessed in the strategic partnerships that some cooperatives have built with the fair trade movement (Satgar and Williams 2008; Wanyama 2008; Develtere et al. 2008).

3.6 Women's Participation in the Third Generation Cooperatives in Africa

A recent study commissioned by the ILO to investigate how women fare in East African cooperatives (Majurin 2012: 39) found that, while women's participation in cooperatives is generally on the rise, in particular due to the expansion of the savings and credit micro-finance institutions (SACCOs sector), women's participation is still lagging behind men's in mixed-sex cooperatives. Factors which negatively affect women's ability to join and effectively participate in cooperatives were found to include the relatively limited access to and control over finance and other physical resources such as land (which is critical for agricultural cooperatives), their low levels of education, as well as cultural perceptions of women's and men's roles in the division of labour in the household. Researchers also found that, despite the fact that the three East African countries have upgraded their cooperative policies and acts in line with the internationally-identified principles and values governing cooperatives, there are still instances in legal provisions which implicitly exclude women. Examples pointed out were laws that stipulate than one member per household is admitted into the cooperative which usually result in women's exclusion and thus lead to men's dominance in cooperative membership. In addition, the continued existence of laws that are clearly exclusionary, such as laws which consider married women as minors or laws which prevent women from inheriting a deceased husband's investments in the cooperatives, continues to pose significant barriers to women advancement.

The study, on the other hand, clearly demonstrated that there are strong positive effects that cooperative membership can and do contribute to women's empowerment. Women who participated in the study pointed out significant economic, social or psychological empowerment effects that they gained through participation in the cooperatives. This clearly has significant implications for development policy and practice, together with other lessons that have been pointed out in this review. There is a need to strengthen the design of support programmes against poverty which are specifically targeted to women. The study, for instance, recommends several actions that can be taken at both the cooperative and the support organisation and policy levels to stimulate women's participation (Majurin 2012: 39). At the cooperative level, actions such as conducting reviews of the extent to which women's membership in cooperatives is hampered can identify emerging good practices and help to disseminate such practices throughout the sector. The introduction of targets for women's participation on cooperative boards may be explored and used where they are suitable. At the support organisation and policy level, researchers recommended that support agencies should capitalise on entry points provided by women's sectoral involvement such as targeting sub-sectors or specific activities within sectors where women are already engaged. Such targeting may include promotion of new social service cooperatives (such as care services for the sick and elderly) which provide services that are of high relevance to women and which are provided by women (Majurin 2012: 41).

3.7 Conclusion and Review Implications

This chapter has reviewed a selection of some topical themes and debates emerging in literature in the contemporary context of the resurgence of the cooperative sector in Africa. This resurgence (or what other writers have referred to as a 'renaissance') is taking place in the aftermath of the liberalisation of economies that occurred with the advent of Structural Adjustment Programmes in all of Africa which were implemented simultaneously with the final round of decolonisation that saw the demise of the Apartheid system in South Africa. Even though it may be tempting to look for theoretical debates in the emerging interdisciplinary themes to write academically rigorous reviews about, it should be acknowledged that to speak of some themes discussed in the above review as theoretical debates may not accurately capture the fact that, at this stage, these cannot be referred to as debates as such. They are merely differing perspectives which may or may not develop into fully fledged debates as more academic interest is generated in the cooperative sector.

Whilst the initial intention of doing the review may have been to identify the theoretical debates (as suggested by the topic of the chapter), the outcome, as this review has attempted to portray, was rather the sounding out of key areas of interest in recent research which, no doubt, point to areas that need further research which may also generate areas of controversy that can stimulate debates going forward. The areas identified in the study conducted by Majee and Hoyt (2010) in the United States such as how different types and sizes of cooperatives are affected in the way they foster creation of social capital by factors such as location, demographic composition, size, years in operation, need to be researched in Africa as they could reveal critical knowledge to assist in the crafting of policies for the cooperative sector. With regard to strategies that states adopt to support the cooperative sector, comparative studies on variations in performance in cooperatives where the government has direct interface with beneficiaries and cooperatives where the government-beneficiary interface is mediated through an intermediary can yield a lot of valuable knowledge.

In all, the review has highlighted that recent research confirms that the resurgence of the cooperative sector in Africa is showing the potential of the sector to actively contribute to the continent's growth and that there are prospects for sustained development and expansion of cooperatives as these institutions are being revived to serve the interests of their members.

References

Allen, E. & Maghimbi, S. (2009). *African cooperatives and the financial crisis*. International Labour Office—CoopAfrica Working Paper No. 3. Dar es Salaam: International Labour Organization.

Ballard, R., & Beesley, A. (2013). Cookie cutter cooperatives in the KwaZulu-Natal school Nutrition programme. *Development Southern Africa, 30*(2), 250–261.

Birchall, J. (2003). *Rediscovering the cooperative advantage: Poverty reduction through self-help*. a Working Paper. Geneva: International Labour Organization (ILO).

Birchall, J. (2004). *Cooperatives and the millennium development goals*. Geneva: International Labour Organization.

Develtere, P., & Pollet, I. (2008). Renaissance of African cooperatives in the 21st century: Lessons from the Field. In P. Develtere, et al. (Eds.), *Cooperating out of poverty: The renaissance of the African cooperative movement*. Geneva: ILO/The World Bank Institute.

Develtere, P., Pollet, I., & Wanyama, F. (2008). *Cooperating out of poverty: The renaissance of the African cooperative movement*. Geneva: ILO/The World Bank Institute.

Hussi, P., Murphy, J., Lindberg, O., & Brenneman, L. (1993). *The development of cooperatives and other rural organizations: The role of the World Bank*. Washington: World Bank.

Majee, W., & Hoyt, A. (2010). Are worker-owned cooperatives the brewing pots for Social capital? *Community Development, 41*(4), 417–430.

Majurin, E. (2012). *How women fare in East African cooperatives: The case of Kenya, Tanzania and Uganda*. ILO Report. Dar es Salaam: ILO.

Mendonça, J. (2008). Surviving on the islands: cooperatives in Cape Verde; chapter 10. In P. Develtere, et al. (Eds.), *Cooperating out of Poverty: The Renaissance of the African Cooperative Movement*. Geneva: ILO/The World Bank Institute.

Munkner, H. (Ed.). (1989). *Comparative study of cooperative law in Africa*. Marburg: Marburg Consulting.
Nyamwasa, J. (2008). Jump-starting the Rwandan cooperative movement, Chapter 11. In P. Develtere, et al. (Eds.), *Cooperating out of Poverty: The Renaissance of the African Cooperative Movement*. Geneva: ILO/The World Bank Institute.
Pollet, I. (2009). *Cooperatives in Africa: The age of reconstruction—synthesis of a survey in nine African countries*. CoopAfrica Working Paper No. 7. Geneva: International Labour Organization.
Portes, A. (1998). Social capital: Its origins and applications in modern sociology. *Annual Review of Sociology, 24*, 1–24.
Porvali, H. (Ed.). (1993). *The development of cooperatives*. Agriculture and Rural Development Series No. 8. Washington: The World Bank.
Satgar, V. (2007). *The State of the South African cooperative sector*. COPAC Discussion Paper. Accessed June 15, 2014 from http://www.copac.org.za/.
Satgar, V., & Williams, M. (2008). *The passion of the People: Successful cooperative experiences*. In Africa COPAC Discussion Paper.
Theron, J. (2008). Cooperatives in South Africa: A movement (re-)emerging, Chapter 12. In P. Develtere, et al. (Eds.), *Cooperating out of Poverty: The Renaissance of the African Cooperative Movement.* Geneva: ILO/The World Bank Institute.
Wanyama, F. (2008). The qualitative and quantitative growth of the cooperative movementin Kenya, Chapter 3. In P. Develtere, et al. (Eds.), *Cooperating out of Poverty: The Renaissance of the African Cooperative Movement*. Geneva: ILO/The World Bank Institute.
Wanyama, F., Develtere, P., & Pollet, I. (2009). *Reinventing the wheel? African cooperatives in a liberalized economic environment*. CoopAfrica Working Paper No. 1. Dar es Salaam: International Labour Organization.

Chapter 4
An Examination of the Roles of Cooperatives in Post-conflict Societies

Dorcas Ettang and Andrew Emmanuel Okem

Abstract Cooperatives are gaining prominence for providing a post-crisis response to communal conflicts, environmental disasters, and financial upheaval. This chapter examines the roles of cooperatives in post-conflict environments in Africa. The chapter explores how cooperatives are useful in post-conflict environments and the challenges and limitations they face in fragile and post-conflict societies. The chapter looks at the relationship between cooperatives and conflict mitigation on the continent within the context of development, peace-building and post-conflict reconstruction. It will draw out practical lessons and examples from within and beyond the African continent. The chapter hopes to contribute to existing discourses on conflict prevention and how to embed this on the African continent. This chapter argues that in limiting the future occurrence of violent conflict, cooperatives bring forth the linkages between development and peacebuilding; two very important factors in preventing future conflict and guaranteeing sustainable peace.

Keywords Communal conflicts · Cooperative · Development · Peacebuilding · Post-conflict reconstruction

4.1 Introduction and Chapter Overview

The primary goal of this chapter is to examine the roles of cooperatives in bringing about sustainable peace and development in post-conflict societies. In doing this, the chapter reviews the principles and values of cooperatives and how these can be applied in peace building initiatives. Pertinent examples of the role of cooperatives

D. Ettang (✉)
School of Social Sciences, University of KwaZulu Natal, Durban, South Africa
e-mail: dorcas.ettang@gmail.com

A.E. Okem
School of Built Environment and Development Studies, University of KwaZulu-Natal, Durban, South Africa

A.E. Okem (ed.), *Theoretical and Empirical Studies on Cooperatives*, SpringerBriefs in Geography, DOI 10.1007/978-3-319-34216-0_4 in poverty alleviation, job creation and the broader process of peace building in post-conflict contexts are presented in this chapter. The chapter concludes with some recommendations on how the roles of cooperatives in post-conflict contexts can be further strengthened and enhanced.

4.2 Development, Peacebuilding and Post-conflict Reconstruction

Broadly speaking, economic development includes "all economic, social, technical and institutional changes that occur jointly to improve living standards" (Emmanuel and Macpherson 2007, p. 19). Based on this definition, development has various dimensions and elements. In one dimension, development is achieved by focusing on job creation because, in many cases, low levels of economic production, unemployment, economic marginalisation and exclusion and high levels of poverty are the underlying causes of conflicts. According to the International Dialogue on Peacebuilding and Statebuilding (2010), consultations with representatives of conflict-ridden countries all found economic growth as an important priority in ensuring that groups benefited from the peace process. It was noted that in doing so, attention should be paid to creating jobs, modernising the agricultural sector and developing the private sector while ensuring that the benefits are felt in rural areas, in poor and highly affected regions, and are used in meeting the needs of returnees and demobilised combatants (International Dialogue on Peacebuilding and Statebuilding 2010, p. 37).

Sustainable development and peace-building are inexplicably linked by the main aim of restoring conflict-ridden societies to one of economic stability, peace, and security. These processes occur in a situation where resources are limited and local capacities to restore the economy after protracted conflict are weak or non-existent. Thus, there has been much talk on the integrated model of both development and peace-building in rebuilding post-conflict societies. As noted in the cases of Burundi, Central African Republic (CAR), Democratic Republic of Congo (DRC), Liberia, and South Sudan, the causes of the conflicts have been linked to "poverty, exclusion from economic opportunities, mismanagement of natural resources, and the lack of economic and financial governance" all of which could affect stability in the short and long-term" (International Dialogue on Peacebuilding and Statebuilding 2010, p. 36).

The use of cooperatives in response to crisis of a social, economic, or political nature has gained prominence in recent years. According to Jaffe (2015), economic and social upheavals are the two leading factors accounting for current interests in cooperatives. It has become an opportunity for economic stability for immigrants, low-income workers and for victims of environmental disasters, violent conflict, and social disorder. In Cuba, worker cooperatives in the agricultural sector constitute about 70 % of the country's arable land and new economic reform goals will see the increase of these co-ops in other industries like hair and beauty, auto repair, transport and technical services (Johnson 2015). In the United States, cooperatives have been used as a 'post crisis development model' and in recognising their significant role, the New York City government voted to invest $1.2 million into the co-op movement by developing worker cooperatives (Jaffe 2015). Another example of the role of cooperatives in peacebuilding is the livelihoods programme set up in Uganda to encourage 'cooperative interaction' between groups shifting them from a history or reputation of violence to one of collaboration on projects like building small dams, joint farming initiatives, and joint restoration of roads and markets (Bayne and Vaux 2013, p. 3). In supporting both peace-building and development efforts, OXFAM, in post-genocide Rwanda, supported community-based development programmes that, among other things, offered poverty reduction grants to communities with the hope of encouraging decentralisation and participatory decision making in order to manage the destructive nature of the conflict (Bayne and Vaux 2013).

Cooperatives have contributed to conflict recovery by promoting and nurturing positive inter-ethnic group relations between previously warring groups in conflicts in Bosnia, East Timor, Lebanon, Macedonia, Mozambique, Nepal and Rwanda (Sentama 2009). While the foregoing outlined useful and very beneficial roles of cooperatives, it is important to examine how these values feature in post-conflict situations and countries that are on the path from protracted conflict to stability and development.

Within the context of development and peace-building, community driven reconstruction focuses on development activities that are centred on building relationships, restoring trust and encouraging cooperation amongst groups (Bayne and Vaux 2013, p. 2). It is important that citizens are involved in setting economic development priorities and developing their capacities to actively benefit from the economic growth and development opportunities that might emerge in post-conflict situations (International Dialogue on Peacebuilding and Statebuilding 2010). In Colombia, there was a recognition by cooperatives of the linkages between peace-building and sustainable development because of the realisation that the latter cannot happen unless there is strong commitment and investment in conflict prevention and sustainable peace (Tulus 2007, p. 114). At this juncture, it is important to examine the link between economic development and peace-building.

Looking at peace-building, peace can be defined as not just the absence of war; its complexity goes beyond that to include a "social order considered to be fair by the majority of the people" (Emmanuel and Macpherson 2007, p. 20). Simply put, peace-building is about halting or preventing violent conflict while also supporting efforts at achieving sustainable peace (IDPS 2010, p. 21). Peace-building and state-building, are "two mutually reinforcing processes aimed at supporting the building of effective, legitimate, accountable and responsive states characterised by healthy state-society relationships and by peaceful relations among communities and with external neighbours" (IDPS 2010, p. 17). Central to peace-building is solidifying social cohesion and reconciliation as relationships have been strained and destroyed due to years of conflict.

The Federation of United Mindanawon Bangsamoro Muslim Women-Multi-Purpose Co-operatives (FUMBMW-MPC) in the Philippines have contributed to peace through their peace-building approaches that focus on "addressing members' economic, social and emotional needs, leadership development, promoting awareness of social issues, fostering appreciation of ethnic cultures while at the same time fostering a sense of nationalism, and policy advocacy" (Bertulfo 2007, p. 111). There is already established evidence of the role of cooperatives and cooperative type structures in peace-building, sustainable post-conflict reconstruction and development efforts (Parnell 2001; Birchall 2004; Bateman 2007 and Emmanuel and MacPherson 2007). Cooperatives have a lot to bring to peace-building efforts due to:

- their experiences in small institutions that contribute to the development of democratic civil society,
- their direct inclusion as units of economic reconstruction that provide a form of enterprise that powerfully motivates the productive energies of their members,
- their instantiation of equality and social justice which suggest the larger, bolder conception of a democratic economy as a goal for peace-building, and
- their contribution to the development of a vibrant civil society (Caspary 2007, p. 343).

In peace-building and reconciliation, the inclusion of individuals and groups in society, and dialogue between groups are all key priorities and goals in any post-conflict or fragile state. Cooperatives are able to address these priorities and goals given that their priority is to meet members' economic, social and cultural needs (International Co-operative Alliance 1995). Borrowing then from Emmanuel and MacPherson (2007), cooperatives achieve the following within the concept of peace-building and development: they are agents of social and economic change, create employment and provide services, enhance group cohesion through bringing groups together through the purchase and sale of products and provision of services, bring groups together under one common agenda and through that improve their understanding and mutual respect to others, and more importantly ensure that they are empowered, and their needs are met in a fair, participatory, just and sustainable manner.

4.3 Cooperatives and Post-conflict Environments

Violent conflicts are a global phenomenon and one that affects the psycho-social, economic, security and political components of any society. Borrowing from Parnell (2001), societies are disrupted from a wide range of crisis events like political transitions, economic and financial crisis, natural disasters, and armed conflicts. As the focus of this chapter, the role of cooperatives as a response particularly in armed conflict is multifaceted. It can help to create individual employment, provide basic social relief, foster dialogue, and integrate victims (refugees and internally displaced persons) and perpetrators including ex-soldiers (Parnell 2001).

Conflicts leave damaging consequences on individuals, communities and groups and these impact on the political, social, economic and cultural dimensions of a society. Cooperatives are one of the many solutions that can address the impact and consequences of conflict in societies. As "an autonomous association of persons united voluntarily to meet their common economic, social, and cultural needs and aspirations through a jointly owned and democratically controlled enterprise" (International Labour Organization 2002; International Co-operative Alliance 1995), the needs and aspirations of cooperatives resonate strongly in post-conflict environments because many of the causes of conflict have to do with economic inequalities, social marginalisation and identities competing for more political rights and power. In addition, cooperatives are "people-centred businesses that can, under the right circumstances, enable people to pool their assets, talents and energies in such a way that they can, collectively, meet their own needs" (Birchall 2004, p. 95). This resonates strongly in post-conflict environments where the state is still in the process of developing its capacity, where many areas remain underdeveloped and are still recovering from the scourges of war, and groups and communities can begin to rebuild slowly.

The various forms of cooperatives meet the needs of the vulnerable, victims and perpetrators in post-conflict situations. Consumer cooperatives provide members access to food and other needs; worker cooperatives provide employment; housing cooperatives provide shelter; credit cooperatives provide microfinance, savings, and credit; and agricultural cooperatives provide support to farmers in growing their crops and livestock and accessing the markets (Birchall 2004).

According to the International Labour Organization' (ILO) Recommendation 193 on cooperative promotion, cooperatives should be encouraged and strengthened based on the "cooperative values of self-help, self-responsibility, democracy, equality, equity and solidarity; as well as ethical values of honesty, openness, social responsibility and caring for others". It also sets forth the cooperative principles of "voluntary and open membership, democratic member control, member economic participation; autonomy and independence; education, training and information; cooperation among cooperatives; and concern for community". These principles are all important ingredients in post-conflict environments in assisting individuals and communities in rebuilding their lives. The values are equally integral to the success of peace-building efforts, the restoration of broken relationships, and are applicable to group norms in spite of one's identity. For example, equality can be defined as a value that emanates from the idea that "each person, irrespective of talent, skill, appearance, race, creed or political belief, possesses an intrinsic value and thus as a human is of no greater value than anyone else" (Sentama 2009, p. 67). The view raised by Sentama (2009, p. 67) is an important element in building social cohesion and reconciliation in war-torn communities on their path to recovery. Based on the principles of cooperatives, cooperatives in post-conflict environments must be guided by the following:

a. Cooperatives should be characterised by open and voluntary memberships where members hold shares and take on responsibilities without discrimination. In a post-conflict environment, it is particularly important that cooperatives at the community level are inclusive and open to all groups regardless of their identities.

Members should also work together and be committed to unity and inclusivity in their words and actions. Trust and solidarity should be built across groups of different ethnic, religious and political orientations as they are brought together under cooperatives.

b. Cooperatives should be democratic where members are actively involved in decision-making and no voices are suppressed in these processes. In situations where there are minority ethnic groups, they should participate actively in decision-making and not be sidelined. Those from majority groups must also allow others from minority groups to participate actively. In post-conflict contexts, it is possible that tensions might still be running high since the causes and impact of conflicts are deep-rooted. Sufficient efforts must be placed on not triggering raw emotions or violent reactions.
c. Concern for the community is integral to the practice of cooperatives which is in line with the importance of bottom-up peace-building efforts. In many post-conflict situations, development and peace-building efforts are focused on major cities while it takes longer for the dividends peace to trickle down local outlying areas. Through the formation of cooperatives, communities at the grassroots can begin their development efforts and not wait for state or external intervention. In Rwanda for instance, most cooperatives and similar associations that emerged post-1994 were spearheaded by the people (Sentama 2009, p. 85). Citing the Ministry of Commerce, Industry, Investment Promotion, Tourism and Cooperatives—MINICOM (2006), Sentama (2009, p. 85) notes that in this case, the local government was not aware that most of these bodies existed.

Cooperatives are of most value when they address the causes of conflicts, particularly conflicts of a protracted nature. Weihe (2004: 4–5) writes that cooperatives can respond to the causes of conflict by increasing a desire by groups for stability, creating a level playing field for groups, building democratic leadership and citizen trust, improving access to resources and improving effective governance processes in relation to political and economic issues. A desire for stability will ensure there is local support for peace-building efforts. Secondly, inclusive processes and programmes will help to build intra-group trust and increase the legitimacy of the state. A working paper by the United Nations Research Institute for Social Development (UNRISD) in 2014 concluded that a major policy lesson for governments involved in post-conflict reconstruction and peace-building efforts was to implement "inclusive agrarian development" (Utting et al. 2014, p. 21). The paper emphasised the importance of a policy that enables and meets the needs for land and employment by disenfranchised and marginalised groups in efforts to prevent any form of resistance, crime, or violent conflict on the part of the latter (Utting et al. 2014).

Post-conflict reconstruction is integral to restoring stability to any conflict situation. This is in line with Weihe's (2004, p. 2) view that the goal of reconstruction is "to get the economy operational and bring stability through re-energised small businesses and rural economies". Reconstruction plays a major role in post-conflict societies. Against this backdrop, the goal of reconstruction is to revitalise the economy and restore security through the strengthening of small businesses and economies, particularly in the rural areas. Cooperatives, informed by their principles and values, are strategically placed to achieve these ends.

Sentama (2009) puts forward the argument that cooperatives are key in post-conflict peacebuilding efforts and should be greatly considered as an alternative to other peacebuilding mechanisms. The argument here is that a cooperative can be seen as "a recruiting vehicle through which post-genocide recovery, social cohesion and reconciliation could be driven" like the case of Rwanda (Sentama 2009, p. 1). The Coffee Cooperatives in Rwanda is a pertinent example of the roles of cooperatives in post-conflict societies. The Rwanda genocide from April to July 1994 remains one of the most ruthless episodes of violence on the continent with approximately 1 million people killed within that space of time. A major outcome of the conflict was the broken relationships and the culture of mistrust, fear, suspicion and hatred which is common in post-conflict societies where groups of different identities are pitted against each other and have committed gross acts of violence to the other. Rwanda provides many lessons for post-conflict peacebuilding, reconstruction efforts and the benefits and values of cooperatives in such a context, especially where there are deep-rooted sentiments between conflicting groups even in a post-conflict phase.

Cooperatives emerged in the country in efforts to build a market rather than a subsistence-based economy. Post-genocide Rwanda left many in difficult conditions as people struggled to produce food and travel long distances to sell their produce (Richardson 2004). A 1500-member cooperative supported by the US Agency for International Development (USAID), became a supplier to the Union Coffee Roasters (UCR) a UK based company that buys coffee beans from countries like Brazil and Ethiopia (Richardson 2004). The cooperative impacted the lives of people who were able to afford food, clothing for their children, hire labourers, get healthcare and build new houses for the family (Richardson 2004). According to Sentama (2009), an examination of two cooperatives in Rwanda (Abahuzamugambi coffee and Peace Basket) show that cooperatives have been useful in restoring relationships between conflicting parties as they have had to put aside previously weak relationships characterised by fear, anger and hatred.

4.4 Benefits of Cooperatives in Post-conflict Contexts

Cooperatives contribute to the promotion of peace and stability by "fostering a strong sense of community, participation, empowerment and inclusion among members and restoring interpersonal relationships and peace" (ILO 2014, p. 14). In times of crisis, cooperatives can have "transformative potential in revitalizing struggling sectors, recovery of crisis-stricken local economies, increasing returns to producers and service providers across value chains…" (ILO 2014, p. 14). Some of the benefits of cooperatives are discussed below with lessons from developed and peaceful economies and examples from across the globe. Cooperatives come with a lot of benefits, all of which are important and critical to ensuring peace. These benefits help to address the root causes of conflict, improve broken relationships and help to address the injustices that groups have faced. Possible benefits associated with the roles of cooperatives in post-conflict African societies include skills development, poverty alleviation, increasing local ownership and commitment, reducing dependence on donor countries and humanitarian agencies, contributing positively to humanitarian and international support, conflict prevention and more importantly, improving optimism about the future. Each of these benefits is briefly explored below.

4.4.1 Skills Development

In a continent with high unemployment rates and a high percentage of unemployed youth, cooperatives provide opportunities to begin small enterprises and improve the skills of individuals. A cooperative such as the Seward Café in Minneapolis, USA, allowed young people to try out different positions at the restaurant while improving their skills in cooking, accounting, with some moving on to open their businesses (Jaffe 2015). Cooperatives have focused on imparting a wide range of skills in computer programming, massage therapy and translation (Jaffe 2015). Cooperatives, like the Northern Virginia Electric Cooperative and the Virginia, Maryland and Delaware Association, have offered scholarships to children of their members (Jett 2015). Other cooperatives such as the Mondragon cooperatives were started by "graduates of a school who taught co-operative principles, along with general education, and practical skills of engineering and business (Caspary 2007, p. 342). The expansion of the cooperatives led to the expansion of its educational arm into a college, a specialised university and a post-graduate centre focused on innovation and research (Caspary 2007, p. 342). In other instances, cooperatives have tried to improve women's skills and knowledge in public speaking (Kiriwadeniya 2007, p. 103). After 10 years of Maoist insurgency in Nepal in 2006, cooperatives have "helped women to survive, manage their livelihood options and look after their families through the provision of credit, counselling and skills development" (ILO 2014, p. 14).

In post-conflict environments, skills development becomes particularly important so that individuals have opportunities to improve their economic livelihood and contribute to the economy after periods of violence. In many instances, conflicts leave countries in a fragile state with a breakdown in systems, institutions and processes. Institutions of learning are destroyed and as an outcome of the protracted nature of conflict, many are not allowed to complete their education. With the ability and opportunity to learn new skills, there is space for citizens to enjoy the benefits of peace. It is of value then that cooperatives provide an avenue for learning and personal growth and development.

4.4.2 Poverty Alleviation

Poverty contributes to conflict and is an outcome of it as well. Thus, poverty alleviation programmes have become central to peace-building efforts. The relationship between cooperatives and poverty alleviation is further established by the focus of the International Cooperative Alliance' (ICA) regional strategy and programme on poverty reduction as a priority in West Africa (Kibora 2007). Cooperatives help in providing solutions to poverty and its causes. In reducing poverty, cooperatives "identify economic opportunities for their members; empower the disadvantaged to defend their interests; provide security to the poor by allowing them to convert individual risk into collective risks; and mediate member access to assets that they utilise to earn a living" (ILO 2014, p. 6). In Rwanda, cooperatives such as the Abahuzamugambi was set up with the aim of improving the socio-economic situation of its members (Sentama 2009, p. 90). While perpetrators and victims of the genocide had negative feelings towards each other, they were motivated to join the cooperative to improve their wellbeing, avoid poverty and be able to meet their basic needs like food, shelter and medication (Sentama 2009, p. 97).

In tackling poverty, cooperatives are a source of employment. In times of crisis, they are useful in "formalizing informal employment and generating employment for women and youth in rural and urban areas" (ILO 2014, p. 14). According to Sierra Leone's Deputy Minister of Trade and Industry, Alie Badara Mansaray, solving youth unemployment could be addressed through various ways one of which was to establish viable cooperative movements across the country, which would also reduce poverty and boost the rural economy (Bangura 2013). Unemployed youths are likely to be involved in crime and are easily recruited for violence. In preventing this occurrence, cooperatives can be explored as a possible solution for youth employment. This is of utmost priority especially as youth make up a large proportion of the African population.

4.4.3 Contribution to Local Ownership and Commitment

The involvement of individuals in cooperatives increases their commitment to the peace-building process in their communities and countries as well. This is due to the vested economic interest that they have in the stability of the country. This commitment to the peace-building process and ensuring its success is strong as conflicts and political violence will impact on the success and economic growth of cooperatives. Members of cooperatives will be more inclined to commit to the peace-building process. The opportunities for wealth accumulation, beneficial financial returns, ownership of a business and having the population involved are many of the benefits that can be expected from a co-op success and are positive indicators for peace-building.

4.4.4 Improved Optimism About the Future

In a post-conflict environment, there are high levels of frustrations and despair about the limitations of the peace process and the possibility of success. As a response, cooperatives are particularly useful for building morale and trust after violent conflict. In a Southern Lebanese village, a women cooperative was successful in reviving local traditional products in efforts to rebuild their societies (ILO 2014). The women cooperatives were formed due to vested interest by the people in the peace-building efforts with the view that it will improve their lives in the long-term. In Rwanda, cooperatives like the Abahuzamugambi coffee cooperative helped to overcome the mutual feelings of suspicion, anger, hatred between perpetrators and victims of the genocide. It also assisted in alleviating the causes of division and improved communication among members of cooperatives. Furthermore, it has resulted in solid inter-group relations such as inter-marriage and deep sentiments of friendship, love and support (Sentama 2009, pp. 101–106). These positive views and sentiments have increased optimism about the future and reduced chances of in-fighting.

4.4.5 Reduced Dependence on Donor Countries and Humanitarian Agencies

While humanitarian agencies and donor organisations are important in contributing to post-conflict reconstruction efforts, it is important that they are not relied upon solely for all economic assistance. Weihe (2004, p. 2) posits that humanitarian and relief organisations tend to "create donor mentalities in which people accept hand-outs rather than self-help assistance". This is detrimental to post-conflict reconstruction efforts and for building the capacity of local actors to self-develop. Cooperatives create an environment for independence and bring a long-term focus to the peace-building process.

4.4.6 Increased Reconciliation and Group Support

According to Ariel Bucardo, a leader of the cooperative movement in Nicaragua, the movement was integral in bridging the alliance between warring groups (Utting et al. 2014). In Rwanda, cooperatives have played a role in bridging the divide between groups. As noted by Sentama (2009), Rwanda cooperatives have brought about positive human relationships by bringing groups together around a common goal that can be achieved when they cooperate. For example, the members of the Abahuzamugambi coffee cooperative included survivors and perpetrators of the genocide (Sentama 2009, p. 90). Prior to the creation of the cooperative, relationships between both groups were marred by anger, fear and hatred resulting in suspicions of each other and a breakdown in communication (Sentama 2009, p. 95). According to Knack and Keefer (1997: 1254), "cooperative norms act as constraints to narrow self-interest, leading individuals to contribute to the provision of public goods of various kinds". In countries with identity-related tensions, cooperatives are useful in bridging the divides among groups. In Rwanda, cooperatives have helped people to look beyond the ethnic divides that fuelled the conflict to jointly create a culture of togetherness which has contributed to reconciliation (Richardson 2004). For those seeking justice after the genocide in Rwanda, cooperatives provided emotional support (Sentama 2009).

4.4.7 Conflict Prevention

Cooperatives can help to prevent conflict as formerly divided groups are now focused on common goals and agendas. In addition, cooperatives provide an environment in which tensions or disputes can be easily resolved and triggers for conflict managed. It also distracts people from the conflict as they are engaged in joint economic ventures. The cooperative movement has been successful in halting "land invasions and brought about the concertation" (Utting et al. 2014, p. 21). In the words of Pedro Huerta, the former leader of the Andrew Castro United Front (FUAC) and the president of the Federation of Small Forest Owners, cooperatives were key in building reconciliation and peace in the countryside between "army combatants, peasant producers, and the counter-revolutionary forces" at a time of high levels of popular resistance (Utting et al. 2014, p. 21). Conflict can be prevented through positive inter-group relations within cooperatives.

4.5 Challenges and Failures of Cooperatives

While this chapter has established the roles and benefits of cooperatives in post-conflict settings like Rwanda, Nicaragua and Sierra Leone, it is important to discuss some of the negative consequences or challenges that emerge from how cooperatives function. These include clashes in views and perspectives, lack of funds and requisite skills, fragile peace processes, corruption, and increased competition. We present a brief overview of each of these in the following sub-section.

4.5.1 Conflicts of Interest

Conflicts of interest could be ideological, cultural, or ethnic in nature. In Brazil, the Canudos Agricultural Production Cooperative Ltd. (COOPAC), while working to collectively be productive and improve members' standard of living, had to deal with understanding the different views and perspectives of members. This entailed "unlearning old habits and learning new ones" (Wittman 2007, p. 140). This is even harder in local groups that have been divided by bitterness, hatred and years of violence. Trust can also be broken from minor conflicts. Theft was a major challenge in the case of diamond cooperatives in Sierra Leone and this was attributed to the absence of trust, an unestablished sense of duty, and the fact that some workers were not paid (Diamond Development Initiative 2008, p. 3). Without a deep commitment to co-operatives by all involved, its chances of success are slim.

4.5.2 Lack of Funds and Skills to Run Cooperatives

The lack of funds to train large numbers of unemployed youth in running and building cooperatives and the absence of requisite skills to actively contribute and engage in a cooperative is a major challenge. This challenge is of huge significance because youth make up a large percentage of the population particularly in Africa and the fact that conflict situations leave them unemployed and uneducated poses a challenge for peace and security. In the case of the diamond cooperative in Sierra Leone, some cooperatives did not have the sufficient skills and knowledge on running cooperatives such that they did not understand "budgets or to record information on the diamonds found…did not understand the basics about cooperatives, how loans would be repaid, and how winnings would be shared" (Diamond Development Initiative 2008, p. 3). As a result of poverty, many members from the artisan diamond cooperatives did not have the willingness and finance to contribute funds to the cooperative. This negatively affected group solidarity and reduced shared ownership and commitment to the process (Diamond Development Initiative 2008, p. 3).

4.5.3 Fragile Peace Processes

Post-conflict countries are in a state of fragility and will require long-term efforts before the dividends of the peacebuilding are felt by the majority of the population. At this early stages of post-conflict environments, trust in government and its capacity are sometimes weak and possible interferences by political leaders threaten the success of peace-building and development efforts. This could pose a challenge to the success and sustainability of cooperatives.

4.5.4 Corruption

As with any peacebuilding process, the corrupt practices of soldiers and politicians can hamper the success of development which can lead a country back to war.

In Rwanda, corrupt government officials at the local level and their negative influences on leaders of cooperatives have negatively affected the progress of cooperatives (Sentama 2009, p. 109). The interference of the state in the affairs of cooperatives was also identified as a major challenge in addition to internal problems like poor leadership and mismanagement of funds (Sentama 2009, p. 109). In Sierra Leone, weak cooperatives from the diamond industry were undemocratic as "executive members were selected according to social status rather than merit; understanding was poor and member relations were not equal or fair" (Diamond Development Initiative 2008, p. 3). In addition to this, diamond cooperatives were similar to the previous traditional systems where "gangs of younger workers were managed by older, more powerful community members" (Diamond Development Initiative 2008, p. 3). Furthermore, corruption of some government officials, officials of the Integrated Diamond Management Program (IDMP)/Peace Diamond Alliance (PDA)[1] and cooperative members led to manipulation and misdirection of funds (Diamond Development Initiative 2008, p. 3).

4.5.5 Competition

While most cooperatives in low and middle-income countries are formed mainly at the local levels and are small in scale, they have to compete with bigger corporations in accessing markets. In West Africa for instance, cooperatives that bring together small-scale farmers, low-income micro-finance savers, or small-scale consumers, are contending with stronger and more diversified competition which requires that they cultivate and practice "creativity, adaptability, competence and professionalism" (Kibora 2007, p. 225). This intense competition affects profit levels which does not bode well for the economic sustenance of these groups.

4.6 Conclusion

More attention needs to be paid on cooperatives and their benefits in post-conflict environments particularly in Africa. The dearth of literature in this field shows that much more needs to be examined on the use of cooperatives in peacebuilding efforts in countries like Burundi, Liberia and other countries and how they have

---

[1]The USAID' contracted Management Systems International (MSI) to manage its US$6.5 million Integrated Diamond Management Program (IDMP) and Its Peace Diamond Alliance (PDA) in Sierra Leone.

been used to address the root causes of conflict. Using cooperatives requires a long-term strategy. Africa's agricultural sectors can benefit from cooperatives as is the case with countries like Cuba where a huge percentage of the agricultural sector is run by cooperatives. In Rwanda for instance, thousands of farmers in cooperatives have benefited from financing and high levels of production (Innocent 2015). A similar success of cooperatives has been noted for Kenya. Worth noting is how to ensure that cooperatives continue to uphold the progress made in peace and development and prevent a relapse into conflict in the future.

Cooperatives should always start small from the bottom up in ensuring that it is inclusive and open to all groups. This is of particular importance if the conflict is between groups polarised on the basis of identity. As the case of Cuba shows, there is a need for support for the bottom-up creation of cooperatives to help people live better lives. While cooperatives are meant to be participatory, democratic and inclusive, it has failed to always do that in post-conflict environments where the resources are provided by donors or from the top. For instance, in Sierra Leone, the beneficiary mentality did not bode well for the artisan diamond cooperatives such that members were not moved from dependence to independence but became more of "a classic top-down aid project" (Diamond Development Initiative 2008, p. 3).

Cooperatives are useful for many things such as reducing inequalities, curbing poverty, and tackling unemployment. Governments must give attention to the benefits of cooperatives in peace-building processes and in ensuring sustainable development. They should support co-ops particularly in rebuilding societies economically and ensuring that there is support and buy-into the peace-building process. This could be done through providing start-up funds for potential businesses through a rigorous, fair and transparent process. However, Birchall (2004) notes that while promoting cooperatives, governments must also guarantee their autonomy. While donors provide funding, these are usually limited in time and in resources. It is, therefore, important for governments to identify alternate sources of funding to help start these cooperatives. Cooperatives also need the skills and knowledge to effectively function and be useful. In post-conflict countries like Burundi and Liberia, Poverty Reduction Strategy Programmes (PRSP) have been developed to offer short, immediate and long-term responses in addressing the root causes of conflict. Cooperatives must continue to be included as a key component of any PRSP. Thus, the United Nations (UN), Donor agencies, bilateral partners, NGOs, and governments must continue to include them in drafting these programmes and in assessing their implementation and success.

Human organisations including cooperatives experience conflicts which emerge from different interests and perspectives on what the needs of the organisation are (Caspary 2007). Cooperatives have been able to address conflict through various ways like "informal consultation, inertia, formal majority vote, consensus decision-making schemes, and managerial edict" among others (Caspary 2007, p. 344). Harnessing these values of cooperatives in post-conflict societies will play an invaluable role in bringing about sustainable peace and development.

References

Bangura, M. (2013). *Sierra Leone: Trade ministry to revamp cooperatives*. Concord Times, 29 August 2013. Retrieved March 04, 2015 from http://www.slconcordtimes.com.

Bateman, M. (2007). Financial Cooperatives for Sustainable Economic and Social Development. *Small Enterprise Development, 18*(1), 37–49.

Bayne, S., & Vaux, T. (2013). *Integrated development and peace building programming: Design, monitoring and evaluation, conflict, crime and violence results initiative*. Department for International Development (DFID). Retrieved July 15, 2015 from http://www.cdacollaborative.org/media/89726/iIntegrated-development-and-peacebuilding-programming.pdf.

Bertulfo, L. Y. (2007). Women Co-ops, and peace work. In J. Emmanuel, & I. MacPherson (Eds.), *Co-operatives and the Pursuit of Peace, British Columbia Institute for Co-operative Studies*. British Columbia: New Rochdale Press.

Birchall, J. (2004). *Co-Operatives and the millennium development goals*. Geneva: International Labour Organization.

Caspary, W. (2007). Co-ops, democracy, conflict resolution and peace: Theoretical resources. In J. Emmanuel, & I. MacPherson (Eds.), *Co-operatives and the Pursuit of Peace, British Columbia Institute for Co-operative Studies*. British Columbia: New Rochdale Press.

Diamond Development Initiative. (2008). *Artisanal diamond cooperatives in Sierra Leone: Success or failures*. Diamond Development Initiative, Policy Brief. Retrieved 15 August 2015 from http://www.ddiglobal.org/BlogPDF/Press%20Release-%20%20Artisanal%20Diamond%20Cooperatives%20in%20Sierra%20Leone.pdf.

Emmanuel, J., & MacPherson, I. (Eds.). (2007). *Co-operatives and the pursuit of peace, British Columbia institute for co-operative studies*. British Columbia: New Rochdale Press.

Innocent, G. (2015). Kagame asks Rwandans to tighten belts, "We shall achieve 2015 goals" news of Rwanda, January 1, 2015. Retrieved January 16, 2015 from http://www.newtimes.co.rw/section/article/2015-01-15/184982/.

International Co-operative Alliance. (1995). *Statement of Cooperative Identity*. Adopted by the General Assembly of the International Co-operative Alliance.

International Dialogue on Peacebuilding and Statebuilding. (2010). *Peace building and state building priorities and challenges: A synthesis of findings from seven multi-stakeholder consultations*. Retrieved August 10, 2015, from http://www.pbsbdialogue.org/media/filer_public/54/bb/54bb413c-f64d-481e-8ad1-d75bed08be34/a_synthesis_of_findings_from_seven_multi-stakeholder_consultations_en.pdf.

International Labour Organization. (2002). *Promotion of cooperatives recommendation (193)*. Adopted at the 90th session of the International Labour Conference (Geneva).

International Labour Organization. (2014). *Cooperatives and the sustainable development goals: A contribution to the post-2015 development debate*. International Labour Organization and International Cooperative Alliance: A Policy Brief.

Jaffe, S. (2015) *Can worker cooperatives alleviate income equality?* Aljazeera America, January 13, 2015. Accessed January 16, 2015. Retrieved January 07, 2015 from http://america.aljazeera.com/articles/2015/1/13/worker-run-cooperatives.html.

Jett, C. (2015). *Two electric cooperatives are accepting applications for college scholarships*. The Freelance-Star, Fredericksburg. Retrieved August 14, 2015 from http://www.fredericksburg.com/business/local_business/two-electric-cooperatives-are-accepting-applications-for-college-scholarships/article_7f882f7f-d82a-5955-85e1-b4d7e89f04dd.html.

Johnson, C. (2015). *How Cuba is using cooperatives to decentralize its economy*. Retrieved January 16, 2015 from http://www.care2.com/causes/how-cuba-is-using-cooperatives-to-decentralize-its-economy.html.

Kibora, A. S. (2007). Peace building through co-operation and self help organizations in West Africa. In J. Emmanuel, & I. MacPherson (Eds.), *Co-Operatives and the Pursuit of Peace, British Columbia Institute for Co-operative Studies*. British Columbia: New Rochdale Press.

Kiriwadeniya, S. (2007). Gender emancipation and peace-building through co-operatives: Going beyond the Myths. In J. Emmanuel, & I. MacPherson (Eds.), *Co-Operatives and the Pursuit of Peace, British Columbia Institute for Co-operative Studies*. British Columbia: New Rochdale Press.

Ministry of Commerce, Industry, Investment Promotion, Tourism and Cooperatives. (2006). *Etude de basesur les cooperatives au Rwanda. Raport d'analyse des résultats préliminaires*. Kigali. Task Force pour la promotion des cooperatives. Retrieved June 14, 2015 from http://www.ifad.org/operations/projects/design/98/rwanda.pdf.

Parnell, E. (2001). *The role of co-operatives and other self-help organizations in Crisis resolution and socio-economic recovery*. Geneva: International Labour Organisation.

Richardson, B. (2004). *Coffee buzz lifts war-torn Rwanda*. BBC News UK. March 10, 2004. Accessed January 27, 2015.

Sentama, E. (2009). *Peacebuilding in post-genocide Rwanda: The role of cooperatives in the restoration of interpersonal relationships*. Ph.D. thesis, Department of Peace and Development Research, School of Global Studies, University of Gothenburg.

Tulus, R. (2007). Gender, co-operatives and peace. In J. Emmanuel, & I. MacPherson (Eds.), *Co-Operatives and the Pursuit of Peace, British Columbia Institute for Co-operative Studies*. British Columbia: New Rochdale Press.

Utting, P., Chamorrow, A., & Bacon, C. (2014). *Post-conflict reconciliation and development in Nicaragua: The role of cooperatives and collective action*. United Nations Research Institute for Social Development (UNRISD) Working Paper 2014-22.

Weihe, T. (2004). Cooperatives in conflict and failed states. U.S. Overseas Cooperative Development Council. Retrieved August 15, 2015, from http://www.uwcc.wisc.edu/info/intl/weihe.pdf.

Wittman, H. (2007). Planting Peace: MST Co-operatives and Agrarian Justice in Brazil. In J. Emmanuel, & I. MacPherson (Eds.), *Co-Operatives and the Pursuit of Peace* (pp. 121–146). British Columbia: New Rochdale Press.

Chapter 5
A Review of the Kenyan Cooperative Movement

Ndwakhulu Tshishonga and Andrew Emmanuel Okem

Abstract The Kenyan cooperative sector has undergone tremendous changes over the years. In the colonial era, the sector moved from being an exclusive white-owned state-dependent organisations to the introduction of black-owned cooperatives to raise a middle class to counter popular uprising. In the post-colonial era, the sector has moved from being state-dependent to an independent era where it now operates as a successful business entity. This chapter provides a historical overview of the Kenyan cooperative sector with the goal of establishing key success/failure indicators thereof. The aim is to identify critical lessons that can be applied in other contexts.

Keywords Cooperatives · Kenyan cooperative · Poverty · Unemployment · Self-help

5.1 Introduction and Chapter Overview

The history of the cooperative movement in Kenya cannot be divorced from the political context in which it emerged and operated. This is because, for many years, the sector was used as an instrument by the state (both the colonial and the post-colonial state) for the attainment of political end. The goal of this chapter is two-fold. Firstly, the chapter presents a brief historical account of the cooperative movement in Kenya. In the second part of the chapter, we glean important lessons that could be drawn from the Kenyan experience with the goal of identifying how these lessons can be applied for the success of cooperatives in other contexts.

N. Tshishonga (✉) · A.E. Okem
School of Built Environment and Development Studies, University of KwaZulu-Natal, Durban, South Africa
e-mail: Tshishonga@ukzn.ac.za

A.E. Okem (ed.), *Theoretical and Empirical Studies on Cooperatives*,
SpringerBriefs in Geography, DOI 10.1007/978-3-319-34216-0_5

5.2 A Historicity of the Kenyan Cooperative Movement

In Kenya, the co-operative movement could be traced back to the British colonial period (Gatheru and Shaw 1998). During the colonial era, the main purpose of setting up cooperatives by the colonial government was to market cash crops produced by smallholders through marketing cooperatives. During this period, the cooperative sector was exclusively reserved for white settlers as evidenced in the 1931 Cooperative Societies Ordinance (Nyagah 2012). The Act led to the establishment of the office of the Register of cooperatives together with a special government department charged with the responsibility for registration, audit, and supervision including the dissolution of cooperative societies (Birchall 1997; Hyden 1973).

The colonial government changed its approach to cooperatives in the early 1940s by encouraging the formation of black-owned cooperatives. This strategy was a tacit attempt by the then government to create an African middle class that could counteract the sustained pressure for independence being put on the colonial government (Gyllström 1991). The instrumentalisation of cooperatives for attaining political end was underpinned by the Swynnerton Plan of 1954.[1] This attempt, however, generated little traction among the black population since there was general apprehension about the colonial government's intentions (Gyllström 1991). Further attempts to encourage the growth of African cooperatives include the restriction of the production and sale of cash crops to cooperatives (Gamba and Komo 2004).

Besides providing services in the marketing of the produce of cooperatives, the colonial government's strategy mediated the emergence of a new generation of African coffee farmers in Central Kenya including other areas in western Kenya–Kisii and Bungoma districts (Hyden 1973). Despite government support, the cooperative sector in colonial Kenya was a dual system. On the one hand, the primary black-owned cooperatives were promoted to cater for smallholder farmers. These cooperatives were economically weak and characterised by high failure rates due to mismanagement and a general lack of understanding of business principles and management of cooperatives (Hyden 1973; Muthuma 2011). On the other hand, white-owned cooperatives whose members were mostly of large-scale farmers retained control of national cooperative unions that promoted their economic interests. National unions such as Kenya Planter Co-operative Union (coffee processing), Kenya Farmers Association (cereals) and Kenya Co-operative Creameries (dairy products) operated on a commercial basis and were economically strong (Muthuma 2011: 78). The duality that characterised the Kenyan cooperative sector persisted until the country attained independence when the focus shifted to enhancing the growth of cooperatives owned by Africans.

---

[1]The Swynnerton Plan was an agricultural policy implemented in colonial Keya to bring about rapid development in the country's agriculture focusing particularly on cash crop production among Kenyans previously excluded from cash crop production. The Plan followed a report of a study by Roger Swynnerton, an official of the Department of Agriculture at the time. The plan, however, was largely seen as a political tool to counter the waning popularity of colonial rule.

The post-colonial era (1963–1979) was marked with a system of cooperatives similar to other African countries. During this period, political leaders implemented cooperatives as the principal strategy for achieving 'African socialism'. According to Hyden (1973: 3), "the co-operative society was to African socialism as the private firm was to capitalism and the public corporation was to communism". In reality, the growth of the cooperative movement in Kenya was boosted by the national slogan 'Harambee' which was a call on every Kenyan to pull together in a spirit of self-help and mutual assistance. The government's commitment to the cooperative sector was effected through the Sessional Paper (No. 10 of 1965) titled "African Socialism and its Application to Planning in Kenya" as well as the Co-operative Societies' Act (CAP 490). The Act presented cooperatives as an instrument of the state and resulted in the extension of financial and non-financial support to the sector and placed the cooperative movement firmly under government control. To provide various forms of technical support to smallholder farmers as well as ensuring that innovative agricultural practices reach grassroots farmers, the government implemented extensive agricultural extension programmes (Muthuma 2011).

In post-colonial Kenya, agriculture was earmarked by the state as the dominant sector of the economy. Against this backdrop, the government committed itself to developing the sector through national farms, cooperatives, companies and individual farms (Muthuma 2011: 79). The focus on agriculture translated into the provision of large-scale extension support to agricultural cooperatives as a means to transforming the sector. The concerted support given to the sector resulted in rapid growth in the number of cooperatives. Between 1963 and 1999, the Kenyan cooperative sector grew by about 14 % (Gamba and Komo 2004).

Like in the colonial period, cooperatives development in the independence era were guided by both political ideology and economic rationale. In Kenya, the cooperative movement was vertically structured along a four pyramid tier system (see Table 5.1). The first tier comprises of primary cooperatives that operate at the

Table 5.1 Kind of co-operative tiers

Kind of co-operative tiers	Nature of co-op & membership
The first tier	Primary co-operatives had membership consisted of individuals within a given locality formed the bottom tier
The second tire	Composed of secondary co-operatives or unions whose membership was restricted to primary co-operatives which also operated at a district level
The third tire	Formed by national co-operative organisations whose membership was drawn from the secondary and primary co-operatives that were not affiliated to the union
The fourth tier	This is the top tier consisted of the apex organisation, the Kenya National Federation of Cooperatives (KNFC) formed in 1964 to become the representative body for the movement

Source Muthuma (2011: 80)

local level. These cooperatives are formed at the local level and comprises of individuals as members. In the second tier are secondary cooperatives which comprises a group of individual cooperative. Primary cooperatives come together to form the third tier of cooperative organisation. The fourth tier is the national apex body that represents all cooperatives in the country.

Although actions of the post-colonial government resulted in an increase in the number of cooperatives, it has been criticised for a number of reasons. Firstly, the approach was top-down and failed to recognise that cooperatives are bottom-up member-owned organisations established to meet members' social, economic and cultural needs (Satgar and Williams 2008). The Cooperatives Societies Act (Act of 1966) as well as the Cooperative Societies Rules (of 1969) both asserted government's intention to control the cooperative sector. Gyllström (1991) notes that the control conferred on government includes:

> The exclusive rights of registration, dissolution and compulsory amalgamation of societies. He [the Commissioner of Cooperative Development] was also given the power to supervise budgets and accounts; to approve remuneration, salary or other payments to officers or members of a society, to approve the hiring and dismissal of graded staff; to dictate a society's mode of organization and activity orientation by prescribing the contents of its by-laws; and to control financial transactions through counter-signature of cheques and other instruments.

The above intention translated into the establishment of government agencies and institutions that provide oversight and management functions over the cooperative sector. The integration of cooperatives into government with bureaucratic inefficiencies that characterised government translated into inefficiencies of the cooperative sector. Effectively, the approach led to the decline of the Kenyan cooperative sector in the medium to long term as evidenced by the decreased contributions of the sector to the country's GDP between 1963 and 1973 despite a marked increase in the number of cooperatives and support institutions (Gamba and Komo 2004; Gyllström 1991). The lack of correlation between financial and non-financial support to the cooperative sector and the successes of the sector, Kenya's rising debt profile and the need for the sector to operate according to internationally recognised principles and values of the cooperative movement resulted in the recommendation that the government ends its control and support of the sector. The advocated reforms were aimed at "restructuring, strengthening and transforming cooperatives into vibrant economic entities that can confront the challenges of wealth creation, employment creation and poverty reduction as private business ventures" (Wanyama et al. 2009: 5). The Renewed Growth and Economic Management of the Economy of 1986 sought to introduce market competition into the economy by removing the practice of price control and the reduction of government participation in the economy through, for example, the privatisation of state-owned enterprises (Gamba and Komo 2004). This was later followed by the amendment of Kenya's Cooperative Ordinance Act in 1997 (Argwings-Kodhek 2004). The introduction of the Session Paper (paper 6 of 1997) further emphasised the independence of cooperatives and limited the role of government to that of the creation of an enabling environment for the cooperative sector.

The introduction of reforms produced both positive and negative outcomes for the Kenyan cooperative sector. On the one hand, the reforms eliminated the instrumentalisation of cooperatives and ensured that cooperatives operated in accordance with internationally recognised principles and values of the cooperative movement. In addition, the liberalisation of the country's economy created new business opportunities for cooperatives as the government withdrew from the provision of certain services (Owango et al. 1998). On the other hand, many cooperatives that previously depended on government collapsed in the absence of government support. One of the reasons for the collapse was their inability to compete once the monopoly and subsidies they previously enjoyed were removed (Muthuma 2011). In addition, most cooperatives lacked management skills to transition from being government dependent to operating as independent entities (Wanyama et al. 2009). This challenge was compounded by other prevailing challenges including administrative mismanagement, theft of cooperative resources, the disintegration of cooperatives and nepotism (Wanyama et al. 2009). The collapse of cooperatives during the transition era had a significant impact on the livelihood of many vulnerable groups.

Despite the challenges that Kenyan cooperative sector has faced over the years, the sector has weathered the storms of the transition to becoming one of Africa's most successful cooperative movement. According to the ICA (2013), cooperatives own 70 % of the coffee market, 76 % of diary and 95 % of cotton, and generate 31 % of national savings and deposits. Most of Nairobi skyscrapers were built by cooperatives (Muthuma 2011: 80). Cooperatives are also prevalent in rural areas employing appropriately 250 000 people. There were approximately 14,126 registered cooperatives in Kenya 60 % of which operated in the financial sector (Nyagah 2012). Financial cooperatives achieved this feat by diversifying their customer base and providing innovative financial products. This led to financial cooperatives recording 65 % growth during the liberalisation era (Muthuma 2011). Institutions such as the Co-operative Bank of Kenya Ltd play an important role in the country's cooperative sector.

In the financial sector, cooperative saving schemes have performed well by amassing substantial savings for their members. Savings and Credit Cooperatives (SACCOs) such as Mwalimu Cooperative Savings & Credit Society Limited, Kenya Bankers Saving and Credit Co-operative Society, Hazina Sacco Society Limited, Unaitas and United Women SACCO are examples of leading and productive SACCOs in Kenya. Estimates in 2011 showed that SACCOs "mobilized deposits and share capital amounting to USD 2.25 billion (Ksh. 189 billion) and loans to members of USD 2.25 billion" (Nyagah 2012: 9). The increase in the capital base of financial cooperatives in Kenya has seen an increase in investor's confidence in the sector. The sector continues to adapt to changing financial practices as evidenced by the recent introduction of e–Kenya which has repositioned CFIs to leverage the benefits of mobile technology in the provision of financial services such as the mobile money platform M-Pesa. CFIs in Kenya have ensured that members have 24 h access to financial services (Timmins 2014).

As primary coffee producing country, coffee producing cooperatives play an active role in the country's economy. The Kenya Cooperative Coffee Exporters Limited is the umbrella body of coffee producing cooperatives that produce about 60 % of Kenya coffee (Kenya Co-operative Coffee Exporters Limited 2014). Through the Kenya Cooperative Coffee Exporters Limited, coffee producers have direct linkage to the international market and better returns on investment (Kenya Co-operative Coffee Exporters Limited 2014).

One of the defining features of the Kenyan cooperative sector is the presence of horizontal and vertical networks among cooperatives. The Cooperative Alliance of Kenya, an apex body representing approximately 14,000 cooperatives, has played an active networking role in driving the agenda of the cooperative movement particularly on policy related issues (The Cooperative Alliance of Kenya Limited 2014). This includes policy advocacy and linkage to support among cooperatives.

The success of the Kenyan cooperative sector has also been linked to the Co-operative University College of Kenya, a Constituent College of Jomo Kenyatta University of Agriculture and Technology, which plays a key role in the provision of education and training on cooperative related issues (Co-operative University College of Kenya 2014). The college provides both degree and non-degree programmes that are geared towards improving understanding and management of cooperatives. The ultimate goal is to produce a crop of managers that are adept with the management of cooperatives.

5.3 Critical Insights from the Historicity of the Kenyan Cooperative Sector

Although the Kenyan cooperative sector has been characterised by successes, Ogina (cited in Pivot 2009: 41) notes that the control of the sector by the government is a contributory factor to the failure of cooperatives. Ogina went on to add that enterprise culture does not come from the government. The failure of state-led cooperatives is attributed to the fact that government's administrative role stifled growth and innovation in the sector. In addition, the financial and non-financial incentives created a dependency syndrome which resulted in the inability of cooperatives to operate independently of government.

The recent resurgence of the country's cooperative sector has been attributed to the liberalisation of the cooperative sector. Rather than being instruments of the state, cooperatives, in the new dispensation are bottom-up autonomous member-owned organisations that are formed to meet members' needs. In doing this, cooperatives have been able to re-establish their principles and their existence aligned to these principles and values.

In addition to the independence of the Kenyan cooperative sector, member education has also been critical to the success of cooperatives. In that regard, the Cooperative College has and continues to play a critical role. The presence of apex cooperative organisation has also been critical in advancing the success of the sector through advocating for friendly cooperative policies. Similarly, integration of cooperatives has enhanced the success of cooperatives as they are able to access resources within the network.

The Kenyan success story was further emphasised by Nyawinda (cited in Pivot 2009) who indicated that the cooperative sector in Kenya has achieved a critical mass of cooperatives and members are able to contribute to enormous funds that come back to them in credit, savings, bonuses and dividends. For instance, Develtere and Pollet (2008) reported that one of the largest cooperatives in Africa in terms of membership is Harambee Savings and Credit Co-operative Society in Kenya with 84,920 members followed by Mwalimu Savings & Credit Co-operatives Society with a membership of 44,400. These results point to a sector that now operates independently of government's financial support. Due to its successes, the Kenyan Model is being benchmarked for replication in many African countries including Lesotho, Malawi, Rwanda and Tanzania (Pivot 2009).

5.4 Conclusion

This chapter aimed to provide a historical overview of the Kenyan cooperative movement. In presenting the historical account of cooperatives, the chapter attempted to highlight factors that contributed to the success and failures of cooperatives in Kenya. The historicity of the Kenyan cooperative sector showed that cooperatives in both colonial and post-colonial eras were used to advance the interest of the state. An outcome of the state-led approach was the failure of the cooperative sector to develop into an independent vibrant movement. The chapter noted that although the transition from the state-led approach to an independent cooperative movement resulted in the failure of many cooperatives, the transition nonetheless led to the emergence of a strong cooperative sector that operates in line with the internationally recognised principles and values of the cooperative movement. The Kenyan cooperative sector thus provides valuable lessons that could be applied in other contexts.

References

Argwings-Kodhek, G. (2004). *Feast and famine: Financial services for rural Kenya*. Tegemeo Institute of Agricultural Policy and Development: Egerton University.

Birchall, J. (1997). *The international co-operative movement*. Manchester: Manchester University Press.

Develtere, P., & Pollet, I. (2008). Renaissance of African co-operative in the 21st century: Lessons from the field. In P. Develtere, I. Pollet & F. Wanyama (Eds.), *Cooperating out of poverty: The renaissance of the African cooperative movement*. Geneva: International Labour Organisation (ILO).

Gamba, P., & Komo, I. (2004). Evolution, growth and decline of the co-operative sector. In *A paper prepared for the centre for governance & development (CGD).*

Gatheru, W., & Shaw, R. (Eds.). (1998). *Our problems, our solutions: An economic and public policy agenda for Kenya*. Nairobi: Institute of Economic Affairs.

Gyllström, B. (1991). *State administered rural change: Agriculturual cooperatives in rural Kenya*. London: Routledge.

Hyden, G. (1973). *Efficiency versus distribution in East African co-operatives*. Nairobi: East African Literature Bureau.

Kenya Co-operative Coffee Exporters Limited (2014). Brief introduction to KCCE. Retrieved June 13, 2015, from, http://www.kencaffee.coop/.

Muthuma, W. (2011). *Economic cooperation in Kenyan credit cooperatives: Exploring the rile of social capital and institutions*. Unpublished Ph.D. thesis submitted to the Faculty of Commerce, Law & Management at the University of Witwatersrand, Johannesburg.

Nyagah, W. N. (2012). *Co-operatives as potential channel for enhancing financial inclusion*. In Paper presented at the enhancing financial innovation and access, Abuja, Nigeria.

Owango, M. J., Staal, S., Kenyanjui, M., Lukuyu, B., Njubi, D., & Thorpe, W. (1998). Dairy co-operatives and policy reform in Kenya: Effects of livestock service and milk market liberalisation. *Food Policy, 23*(2), 173–185.

Pivot (n.d.) (2009). Driving growth in KwaZulu-Natal, KwaZulu-Natal official publication of the ministry of economic development and tourism. Pietermaritzburg: KZN-Ministry of Economic Development and Tourism.

Satgar, V., & Williams, M. (2008). The Heiveld cooperative. In The passion of the people: Successful cooperative experiences from Africa: COPAC research report. Retrieved May June 10, 2013 from, www.copac.org.za.

The Co-operative University College of Kenya (2014). *About us*. Accessed December 31, 2014, from http://cuck.ac.ke/?page_id=5086.

Wanyama, F., Develtere, P., & Pollet, I. (2009). *Reinventing the wheel? African cooperatives in a liberalized economic environment.* Coop AFRICA Working Paper No.1. International Labour Office. Cooperative Programme (EMP/COOP). Geneva, Switzerland.

Chapter 6
Coffee Cooperatives in Cameroon: An Assessment of the Causes of Coffee Cooperatives' Collapse in the Post-adjustment Period

Gillo Momo-Lekane

Abstract Coffee cooperative associations were the most important channels through which coffee farmers used to sell their commodities as well as receiving various forms of government assistance before the adoption of the market liberalisation policies in Cameroon. These forms of support included extension services, subsidies and in some cases social services such as healthcare and education. With the adoption of the liberalisation policies, the coffee commodity chain shifted from a chain regulated by the heavy hand of the state to the current one regulated by market forces. One of the key consequences was that this shift contributed to the collapse of the majority of coffee cooperatives. This chapter investigates the factors that contributed to the fall of the coffee cooperatives in the post liberalisation period in Cameroon. It argues that although the proximate causes of the collapse of the agricultural cooperatives can be attributed to the country's macroeconomic policy shift from pre-adjustment to the post-adjustment period, the nature of the cooperative system as 'government institutions' in the pre-liberalised economy was its original sin that predisposed newly established cooperatives to an eventual collapse. As such, it is understandable that coffee cooperatives fell with government withdrawal from the economy after the country's adoption of the IMF- and World Bank-imposed neoliberal macroeconomic policies.

Keywords Cooperatives · Coffee cooperatives · Market liberalisation · Cameroon

G. Momo-Lekane (✉)
Worker's College, Durban, South Africa
e-mail: momogillo@gmail.com

A.E. Okem (ed.), *Theoretical and Empirical Studies on Cooperatives*,
SpringerBriefs in Geography, DOI 10.1007/978-3-319-34216-0_6

6.1 Introduction

It is estimated that about half of the world's population makes a living through cooperative enterprises. Since the colonial period, cooperatives have been promoted in almost all African countries as critical economic development institutions because of the commonly accepted assumption that they are a crucial means for poverty alleviation and improved livelihoods for many households. In Cameroon, coffee cooperative associations were the most important channels through which coffee farmers used to sell their commodities as well as receiving various forms of government assistance in the pre-liberalisation era. With the adoption of the liberalisation policies the coffee commodity chain shifted from a chain regulated by the heavy hand of the state to the current one regulated by market forces. One of the immediate consequences of this shift was the collapse of the majority of coffee cooperatives. The aim of this chapter is to investigate the factors that contributed to the fall of the cooperatives in Cameroon's liberalised economy. Data was collected in December and January 2013 through qualitative interviews of purposively selected coffee farmers and site visits in Dschang, Bafousam, and Santchou. An in-depth interview was conducted with an official of the central cooperative (UCCAO[1]) in Bafoussam. The chapter starts with a broad literature review on the nature and role of cooperative associations. Then compares and contrasts coffee cooperatives before and after the market liberalisation policies, while assessing the factors that contributed to the fall of the coffee cooperative systems.

6.2 Paradigm Shift: From Community-Owned and Managed Institutions to Government Institutions

The traditional conceptualisation of cooperatives contrasts greatly with how they originated and were managed in many countries in Africa. The history of cooperatives in Africa is long and complicated. The roots of modern cooperatives can be traced back to Africa's colonial past which resulted in diverse colonial traditions of cooperation on the continent (Jara and Satgar 2008). The colonial masters used cooperatives as instruments to organise agricultural production and mainly the cash crop sectors according to the trade needs of their home countries. As such, cooperatives were first and foremost a creation of the colonial masters. Colonial-era cooperatives in Africa were used as instruments to organise cash crop sectors that

---

[1]The French acronym UCCAO stands for *Union Centrale des Coopératives Agricole de L'Ouest*translated as the central cooperative for agricultural cooperatives of the Western Region. Headquartered in Bafoussam the capital city of the Western Region of Cameroon, UCCAO is the umbrella body for mainly coffee cooperatives that is comprised of six major cooperatives (CAPLAMI, CAPLAME, CAPLAMBAM, CAPLANOUN, CAPLAHN, and CAPLANDE) located in the main town in the Mifi, Menoua, Bamboutos, Noun, Haut Nkam, and Nde Divisions.

served the imperial and capitalist needs of the colonisers. In French colonies, such as Cameroon they were created as parastatal rather than independent bodies. They were subsequently inherited by the newly independent governments who used them both as development tools to deliver services to people and as political tools to exert some form of control over people and communities. In so doing, cooperatives became devoid of their essential characteristic and became mere top-down, inorganic and hybrid institutions. They were government-funded instead of member-funded and lacked the democratic principle which was at the heart of the traditional cooperative system. It can then be argued that at birth, the seeds of self-destruction were already planted in the newly created cooperatives in Cameroon.

6.3 The Rise and Fall of Coffee Cooperatives in Cameroon

The history of agricultural cooperatives in Cameroon can be traced back to the country's colonial period. According to Champaud (1969), '*pre*-cooperative' societies already existed in the Western Region of Cameroon before the introduction of modern cooperatives. They were created by the *Société Africaine de Prévoyance* (African Provident Society) which pioneered farmers' organisations. The aim was the improvement of farming methods and the increase of production. The earliest 'pre-cooperatives' were created in Dschang and Foumban in 1922 and 1933, respectively, and can be seen as the precursors of the cooperative system in the country. These grandparents of cooperatives planted the seeds of the many primary cooperatives that existed in the western province of Cameroon before independence and were later grouped into seven secondary cooperatives. Two were based in Dschang, the most important centres for Arabica and Robusta coffee production in the region. The other five were based in Bafoussam, Foumban, Mbouda, Banganté and Bafang. In 1959, these seven cooperatives organised themselves and created one central union, the *Union Centrale des Coopératives Agricole de L'Ouest* (UCCAO) with each of them retaining their relative autonomy (Champaud 1969).

The newly formed UCCAO's task was essentially to commercialise coffee while the role of the seven departmental cooperatives was to ensure the collection and processing of coffee beans. In addition, these cooperatives were responsible for assisting farmers to improve their farming methods. Champaud (1969) explains that coffee farmers' participation in cooperatives was organised through traditional leaders (the chiefs and some of their notables), or through the seven to nine delegates who were elected members of the cooperative's administrative committee. In turn, three members of each administrative committee were chosen to sit on UCCAO's administrative council.

Cooperatives, which became the primary administrative structures for coffee farming and processing after independence, were involved in all aspects of the coffee chain including collecting, storing, hulling, grading, extension services and financing (World Bank 2010). In 1961, the government granted UCCAO the monopoly to commercialise Arabica in the West Region. This monopoly gave UCCAO effective control over most of the Arabica coffee farmed in this jurisdiction. In line with the principles and values of cooperatives, UCCAO's and other central unions' mandate was not only the commercialisation of coffee but also included a developmental role geared towards improving its members' welfare, including social services such as healthcare and education.

It is usually argued that cooperatives are created when the market fails to provide the required goods and services at affordable prices and acceptable quality (NCBA 2005 cited in Ortmann and King 2007). Agricultural cooperatives are therefore of great importance for rural farmers because they can empower farmers by exposing them to economic opportunities that improve their livelihoods. UCCAO's official interviewed in Bafoussam explained that coffee cooperatives were created to facilitate the commercialisation of coffee both for the producers and the government. Cooperatives in Cameroon were not created as a response to market failure as postulated above. Rather, they emerged as a channel through which the government could facilitate the cultivation and commercialisation of coffee.

It was clearly stated in Cameroon's cooperative bylaws that cooperatives have several functions and goals including, amongst others, "collecting, processing, sorting, sacking, storing, and marketing primary products; promoting savings and capital formation; supplying agricultural inputs; providing training and cultural activities; defending the interest of the membership; and improving social conditions" (Gellar et al. 1990: p. 34). This conception of the cooperative in Cameroonian law shows that cooperatives were designed to be a developmental institution operating for the welfare of farmers rather than profit making. As developmental institutions, the government offered cooperatives loans, subsidies for fertilisers, pesticides and fungicides; funding for agricultural extension workers; and grants for equipment, buildings, and vehicles. Within this framework, the UCCAO functioned not only as a cooperative union but also as the main developmental institution, often providing healthcare, education, and various other public goods to farmers via its departmental cooperatives based within each farming community (Gellar et al. 1990).

In the Western Region of Cameroon where the research was conducted, it was noted that cooperatives were divided into two main categories: the primary or cooperative union and the secondary cooperatives or departmental cooperatives with their representative bodies based at the departmental level (examples are CAPLAMI and CAPLAME for the Mifi and the Menoua Divisions, respectively). Primary cooperatives are the central cooperatives based at the regional level (examples here are the UCCAO based in Bafoussam, and the NWCA which is a similar structure based in Bamenda, the capital of the North West Region). These primary cooperatives are considered as mother cooperatives. They manage and regulate the secondary cooperatives comprising numerous sub-cooperatives, which can be seen as third level cooperatives, located in each farming community.

For coffee farmers, the language of the 'collapse of cooperatives' used by many interviewees refers to the closure and abandonment of these numerous cooperatives located in close proximity to the various farming communities, allowing them to receive extension services, collect agricultural inputs and sell their beans, among other activities. It could be argued that these cooperatives were the life blood, the *raison d'être*, of both secondary and primary cooperatives. The 'collapse of cooperatives' discourse speaks both to the struggle and fall of the above explained cooperative system with its accompanying severe consequences on farming households who depended heavily on them for their survival.

6.4 Collapse of Cooperatives as Collateral Damage of the Liberalisation Policies

The cooperative system collapsed after liberalisation due to the withdrawal of government support, mismanagement, falling revenue, maladministration, and increased competition from the private sector (World Bank 2010). As discussed above, the cause of the collapse of the cooperative system is commonly attributed to the withdrawal of government support. During field visits to UCCAO, CAPLAMI, and CAPLAME, it was observed that buildings and sections of buildings which had been traditionally used exclusively for the management of coffee farming activities were sometimes being used for purposes other than coffee farming– if they were not totally abandoned. At the two CAPLAME sites in the city of Dschang, the researcher noticed that one of the buildings was being used as a chicken farm, others as sites for social gatherings such as places of worship, for functions, and sale of food crops such as maize. At both CAPLAMI and CAPLAME, it was observed that a number of pieces of machinery and trucks which had been used to process and transport coffee from remote cooperatives were in disrepair and consequently had been abandoned. In Bafoussam, it was noted that UCCAO was renting out sections of its building as office space to private companies. This was in sharp contrast to the pre-liberalisation period when these structures were solely and exclusively used by UCCAO and secondary cooperatives for the management of coffee production.

In a conversation with one of the officials at the central cooperative in Bafoussam, it was highlighted that despite the challenges posed by the market liberalisation policies which had greatly affected the size and scope of its activities, UCCAO remains the main player in what was left of the Arabica market. This is consistent with the World Bank study which shows that UCCAO exported 41 % of Cameroon's Arabica coffee during the 2008–2009 season (World Bank 2010). It was also discovered that the government was trying to revitalise the coffee sector as one of its key priorities. One of the strategies used was the free distribution of seedlings to farmers through UCCAO and its six secondary cooperatives. However, farmers were very sceptical about this new initiative because they were still recovering from the trauma of the fall of the coffee sector which had obliged them in the process to adopt new coping mechanisms mainly based on food crop farming. Although these activities conducted by UCCAO allowed the institution to keep afloat, it was clear that in the post-liberalisation period, UCCAO stands as a shadow of a vibrant organisation that it had been in the pre-liberalisation period in its various aspects including capital, assets, human resources and reputation, to cite but a few.

6.5 Original Sin of the Cooperative System

The distinction between the so-called vibrant and successful cooperatives in the pre-liberalised economy vis-à-vis the collapse of cooperatives and their struggle to survive in the post-liberalisation period, usually leads to a conclusion that the neoliberal policies adopted by African countries are to be blamed for the fall of the agricultural cooperatives in Cameroon and on the continent. This cause-and-effect explanation only accounts for the proximate causes and not the root causes that could best explain why many cooperatives did not survive or are struggling to survive in a neoliberal age. This begs the question of why many agricultural cooperatives were unable to survive after the macroeconomic shift brought about by the structural adjustment policies of the late 1980s. The case of coffee cooperatives in Cameroon as illustrated provides an answer to this question.

Cooperative systems were flawed from their conception and inception by the colonial masters in farming communities in Cameroon. The newly independent government inherited a legacy of using cooperatives as government entities for development and political control. The roots causes of their demise can then be located in the manner in which they were conceived and established as the antithesis of what normally should have been the essence and nature of cooperatives, which are community-based and member-owned institutions operating independently for the benefit of their members. Instead, in Cameroon, cooperatives were the principal agent mandated by the National Marketing Boards, ONCPB (*Office National de Commercialisation des Produits de Base*) to buy coffee from farmers for export.

Coffee farmers then had a strong incentive to join a cooperative as the only way to access benefits provided by the government through the Marketing Board. The Marketing Board also played an important role in stabilising the highly volatile coffee prices to ensure that farmers were protected against the boom and bust of international coffee prices. The government also used cooperatives as channels to provide agricultural credit schemes, extension services, and subsidies for fertilisers, pesticides, and other key inputs. Coffee farmers were implicitly obliged to join cooperatives otherwise they would have been unable to receive benefits such as key inputs, specialised support services, or even the commercialization of their products. It was highlighted during the researcher's conversation with coffee farmers in Dschang, Santchou and Bafoussam that cooperatives were at the centre of many rural communities' lives as providers of essential services such as access to primary healthcare and education. Some of these services were located within the cooperative's precincts. The human resources employed in the cooperative system were on the government payroll. It was also noted that, later, cooperatives were used not just for the promotion and enhancement of coffee farmers' livelihoods but were also used to achieve strategic political goals by rewarding political supporters and their families with employment (Muenkner and Shah 1993). By owning cooperatives and using them to achieve political goals, the government removed these institutions' voluntary and independent character that goes hand in glove with the principle of democratic member control. Thousands of cooperative workers who were civil servants were retrenched with the market liberation policies and the consequences were dire for their households.

6.6 Conclusion

The aim of this chapter was to assess the causes of the collapse of coffee cooperatives in the post-adjustment period in Cameroon. It was observed that the original sin of the cooperatives system in Cameroon started with their creation by the colonial powers as tools to facilitate the production of cash crops which served to achieve the trading goals and the imperialist project of the colonial home country. The legacy of this misconception of the aim and objectives of the essence of the cooperatives was carried over to newly independent African countries where cooperatives were then used as developmental and political control tools. With the advent of the structural adjustment programmes, cooperatives suffered a severe blow as governments were asked to withdraw from the economy and to abide by the principles of the free market. Senegal, for instance, had a large Ministry for Cooperative which was forced to shut down by the World Bank. At the same time, other support services that were attached to cooperatives were privatised (Satgar and William 2008). As governments withdrew from economic activities, cooperatives were forced to operate on their own. This was an impossible task given that these institutions were government institutions by design and were not private institutions. As such, the majority of their employees were government employees. With the heavy and visible hand of the government out of the coffee sector, what followed was the inevitable natural collapse of the cooperative system as they were not meant to operate as private entities, let alone compete in an increasingly globalised economy. Cooperatives that managed to survive have been struggling to stay afloat. Satgar and William (2008) observe that those that are still surviving often do so because of the passion and determination of their members who strive against all odds as they are faced with issues of inequalities in the world trade system. For cooperatives to be successful and to achieve their goal of serving their members equitably, they should not be imposed on people and communities but be allowed to develop organically from the needs and concerns of their members. The role of the government should be limited to encouraging communities and groups to start cooperatives and most importantly to own them without any external interference.

References

Champaud, J. (1969). Cooperatives et development: L'U.C.C.A.O. *Les Cahiers D'Outre-Mer, Tome XXII, 55*, 95–100.

Gellar, S., Oakerson, R. & Wynne, S. (1990). *An institutional analysis of the production, processing, and marketing of arabica coffee in the west and north west province of cameroon*, Decentralization, Finance and Management Project (DFM).

Jara, M. K., & Satgar, V. (2008). *International cooperative experiences and lessons for the eastern cape cooperative development strategy: A literature review*, Cooperative and Policy Alternative Centre, Eastern Cape Socio Economic Consultative Council (ECSECC) Working Paper No. 2.

Muenkner, H. H., & Shah, A. (1993). *Creating a favourable climate and conditions for cooperative development in Africa*. Geneva: International Labour Organization.

Ortmann, G. F., & King, R. P. (2007). Agricultural cooperatives I: History. Theory and problems. *Agrekon, 46*(1), 40–68.

Satgar, V. & William, M. (2008). The Passion of the People: Successful Cooperative experiences in Africa. Johannesbourg: COPAC.

World Bank (2010). *Cameroon coffee supply chain risk assessment*, Draft report, All ACP agricultural commodities programme.

Chapter 7
Cooperatives in Vogue: A Case Study of Cooperatives in South Africa

Meron A. Okbandrias and Andrew Emmanuel Okem

Abstract The 2005 Cooperative Act and the subsequent amendments and other related policies and support have invigorated the cooperative sector in South Africa. In fact, if one looks at the number of cooperatives registered in South Africa, it is quite impressive. The legislative environment and the financial support that the country provides have resulted in increased number of newly registered cooperatives. There were about 4000 cooperatives in 2005 compared to the 50,000 currently registered cooperatives in South Africa. Compared to population growth and the rate of urbanisation in South Africa, this number is very high. Despite the impressive growth in the number of newly registered cooperatives, the sustainability of many of the registered cooperatives is doubtful. This chapter provides an overview of the history of the country's cooperative sector with the goal of identifying factors that contribute to the high failure rates of cooperatives. Based on the available literature and statistics, the chapter shows that financial incentives and the creation of government support institutions are not enough to foster sustainable and successful cooperatives.

Keywords Apartheid · Cooperative · Economic transformation · Sustainability · Cooperative incentive scheme

M.A. Okbandrias (✉)
Department of Public Administration, School of Government,
University of the Western Cape, Cape Town, South Africa
e-mail: okbandrias@uwc.ac.za

A.E. Okem
School of Built Environment and Development Studies,
University of KwaZulu-Natal, Durban, South Africa
e-mail: okem@ukzn.ac.za

A.E. Okem (ed.), *Theoretical and Empirical Studies on Cooperatives*,
SpringerBriefs in Geography, DOI 10.1007/978-3-319-34216-0_7

7.1 Introduction and Chapter Overview

Cooperatives are characterised by mutual help, workers solidarity, and social responsibility. Even though governments and countries support cooperatives in different forms, it is indeed the community that is the organic source of such an organisation. The question that this chapter touches is whether the South African socio-economic and political context is conducive for the organic emergence and growth of the cooperative sector. The chapter argues that the current climate seems to have failed to glean lessons about past experiments with the cooperative movement. This chapter is divided into three sections. In Sect. 7.2, we explore the South African cooperative in pre-democratic South Africa. The review focuses on how the sector emerges as well as its relationship to the state. In Sect. 7.3, we examine the cooperative sector since the transition from the Apartheid system. In the review, we focus on the contradiction between the recent upsurge of cooperatives in terms of registration and the high failure rates. In this part of the chapter, we review factors that contribute to the failure of cooperatives and end the chapter with recommendations on how to address the identified challenges.

7.2 A Historical Account of Cooperatives in Pre-democratic South Africa

Cooperatives have a long history in South Africa and this history has always been intertwined with the state. According to the Department of Trade and Industry (DTI) (2012: 36), the cooperative sector is an economic vehicle in South Africa that can be traced back to the early 19th century. Co-operatives began with the introduction of white agricultural cooperatives which aimed at developing and building the white farming community. Satgar (2007) notes that South Africa's first cooperative, a consumer cooperative, was established in Pietermaritzburg in 1892. However, it was not until 1908 that the country's first cooperative legislation (the Co-operative Societies Act) was passed in the Transvaal (Arando et al. 2010).

Under the colonial and Apartheid regimes, cooperatives were used mainly to advance the economic interests of white South Africans most of whom operated in the agricultural sector. The state's desire to promote white-owned cooperatives is evident in the passage of various legislations and the establishment of cooperative support institutions (such as the Co-operative Societies Act enacted in 1908 in the Transvaal, the Land and Agricultural Bank [the Land Bank], the Co-operatives Societies Act[Act 28 of 1922], the Co-operative Societies Amendment Act[Act 38 of 1925], the Marketing Act of 1937[later amended as Act 59 of 1968] and the Co-operatives Act[Act 91 of 1981]). As noted above, most cooperatives in colonial and Apartheid South Africa were agricultural cooperatives and by implication, favourable agricultural policies were important for their advancement.

Reflecting on the impacts of the above policies on white farmers, Satgar (2007: 2) argues that "Afrikaner empowerment in the 20th century did not just happen through the logic of capital accumulation but had to also coexist with a logic that met human needs through cooperative forms of organising production and consumption; albeit underpinned by perverse and racialised relations of productions". Effectively, cooperatives were used as a tool for transferring state resources as well as creating white monopoly in the agricultural sector. Such an approach contributed to the growth of agricultural cooperatives in pre-democratic South Africa. The subsidisation of agriculture resulted in the transfer of state finance to white farmers (most were involved in forming and operating cooperatives). The establishment of the marketing board granted white farmers monopoly over the agricultural sector. The monopoly created by the hegemony of white-owned agricultural cooperatives skewed prices of agricultural produce and farm inputs in their favour (Nilsson 1999).

As noted earlier, the primary focus of pre-democratic state was the advancement of white-owned cooperatives (Westoby 2014). Where the state took steps to encourage black cooperatives, this was only to advance the Apartheid government policies for the former homelands. Where black-owned cooperatives were not stooges of the then government, the government took conscious steps towards stifling their growth (Nilsson et al. 2009; Department of Trade and Industry 2004b). Although the Apartheid government proactively attempted to stifle the growth of cooperatives among the black population, there were strong informal financial cooperatives such as the Savings and Credit Cooperatives (SACCOs) also known as *stokvels*. With about 10 million members, *stokvels* constitute an important feature of the South African economy despite operating largely in the informal economy (Department of Trade and Industry 2012). Despite this success, the majority of black-owned cooperatives remained weak and underdeveloped due to government's restrictions with majority eventually collapsing.

During the1980s, cooperatives began to receive recognition from civil society as an instrument for alleviating economic and social deprivation among the very poor of South Africa. For instance:

> Trade unions started developing co-operatives as a vehicle to address retrenched and redundant workers' economic needs. The Union of Metal Workers of South Africa (NUMSA) established the Sarmcol Workers Co-operative (SAWCO), which later collapsed due to lack of knowledge on co-operative governance and management. Emulating NUMSA, the National Union of Mineworkers (NUM) also tried to establish cooperatives in the late 1980s to address the unemployment of mine workers but unfortunately this ended in failure. Churches and community organizations also provided funding independently, for worker co-operatives in response to retrenchments and unemployment during the same period (DTI 2012).

From the foregoing, it can be concluded that the cooperative sector before 1994 developed along two distinct paths. The first path led to the establishment of strong white-owned cooperatives operating in the formal economy. The second path was characterised by the stifling of black-owned cooperatives, resulting in black-owned cooperatives operating largely in the informal sector. In addition, cooperatives in pre-democratic South Africa were deeply linked to the state and their successes were largely dependent on extended state support. Against this backdrop, the cooperative sector was not operating in line with the internationally recognised principles and values of the cooperative movement.

The adoption of the neo-liberal policy in the 1980s led to the deregulation of the agricultural sector where cooperatives mainly operated in pre-democratic South Africa. The deregulation ended state subsidies and attempted to reduce the dependence of cooperatives on government. In the absence of state support and monopoly of marketing boards, many agricultural cooperatives converted into investor-owned firms in the late 1980s and early 1990 (Van Wyk 2014).

7.3 A Historical Account of Cooperatives in Post-apartheid South Africa

South Africa witnessed a paradigm shift in its political system when the African National Congress (ANC) took over the reins of power in 1994. One of the immediate preoccupation of the ANC led government was to address the injustices of the past through radical transformation of the country's unjust economic system created by the Apartheid regime. The government identified cooperatives as one of the mechanisms for poverty alleviation, income generation and bridging the country's economic gap between the rich and the poor (Department of Trade and Industry 2004b). Besides being a tool for rectifying injustices of the past, cooperatives were also conceived as a tool for achieving local economic development (Kanyane 2009). To achieve the goal of economic transformation through the cooperative model, government institutions at the national, provincial, and local government levels were tasked with various roles in promoting the cooperative sector.

Against the backdrop of the racially skewed outlook and outcomes of previous policies, the democratic government focused on the growth of black-owned cooperatives (Department of Trade and Industry 2004a). The government's desire to establish black-owned cooperatives is reflected in the 2004 Co-operative Development Strategy for South Africa, the 2004 Co-operative Development Policy for South Africa, the Co-operative Act (Act 14 of 2005), the Co-operative Banks Act (Act 40 of 2007) and Co-operatives Amendment Act (No. 6 of 2013). In addition to the policies and Acts, the government also institutionalised support and funding for cooperatives including the Cooperative Incentive Scheme (CIS).

The Department of Trade and Industry (2013: 36) defines the cooperative incentive scheme (CIS) "as a 100 % grant for registered primary co-operatives (primary co-operative consists of five or more members)". The objectives of the CIS, as outlined by DTI (2013: 36), are to "improve the viability and competitiveness of co-operative enterprises by lowering their cost of doing business through an incentive that supports Broad-Based Black Economic Empowerment".

Only registered cooperatives, as contemplated in the 2005 Cooperative and as amended by the 2013 Co-operatives Amendment Act, are eligible to be funded under the CIS (Republic of South Africa 2013). In line with the policy thrust of the democratic government, the support offered to cooperatives is biased towards previously disadvantaged groups and women.

Besides direct financial support, the DTI also support cooperatives in terms of training, compliance, administration and drawing up of business plans. There are government departments, agencies, and organisations besides the DTI that support cooperatives by providing financial and non-financial help. An immediate positive effect of the new policy environment, as well as financial and non-financial support of cooperatives, was an upsurge of newly established cooperatives as shown in Fig. 7.1. The Companies and Intellectual Property Commission (2014: 105) acknowledges the correlation between increased government support and increase in the registration of new cooperatives noting that the increase is attributed "to government departments promoting the registration of cooperatives as a vehicle for poverty alleviation and assisting rural communities to grow economically".

Despite the increase in the number of newly registered cooperatives, the strength of the cooperative sector has remained largely unchanged (Okem 2016). This is evident in the high failure of newly registered cooperatives. The Department of Trade and Industry (2010: 9) notes that the country's cooperative sector is characterised by an 88 % failure rate. In 2009 alone, 5415 out of 6089 registered cooperatives in food and agriculture sector failed. That is a failure rate of 89 %. The national picture in terms of the number of dead and surviving cooperatives is presented in Fig. 7.2. In addition to high failure, the country's cooperative sector is also characterised by weak cooperatives that have achieved very little in terms of employment creation and income generation for members (Department of Trade and Industry 2009).

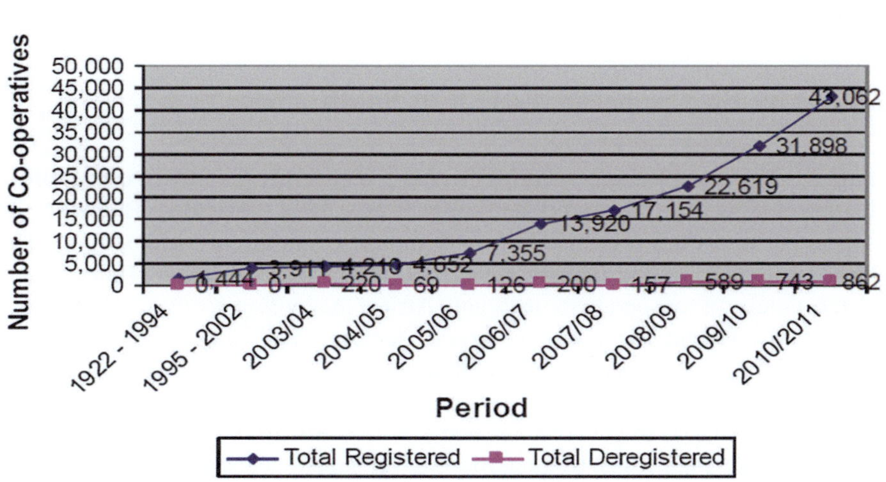

Fig. 7.1 Trends in co-operative registrations and de-registration. *Source* Department of Trade and Industry (2012: 29)

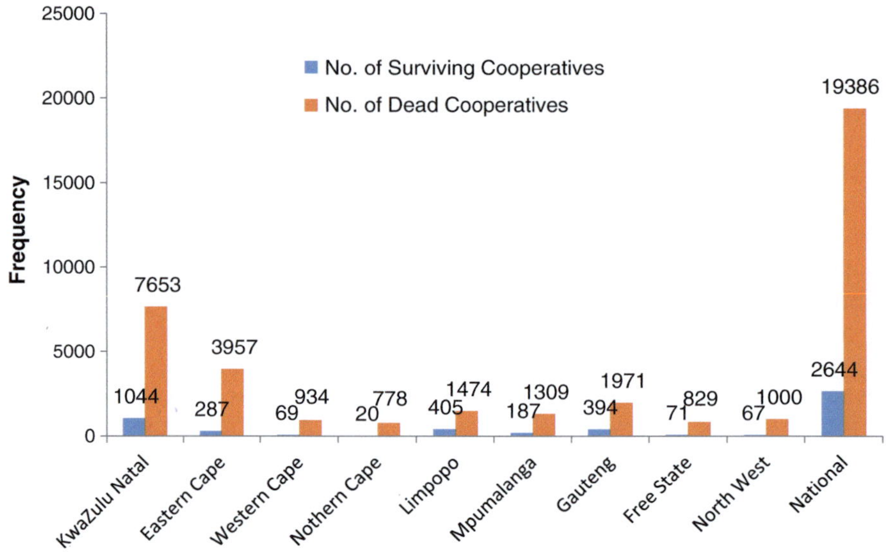

Fig. 7.2 Provincial breakdown and status of cooperatives. *Source* Adapted from Department of Trade and Industry (2010: 9)

Furthermore, the country's cooperative sector is characterised mainly by primary agricultural cooperatives. The cooperative sector also has an insignificant presence in the financial sector. The none participation of cooperatives in financial services has placed restrictions in terms of access to finance for cooperatives as most small emerging cooperatives are not able to access funds from commercial financial institutions.

7.4 Why Do Cooperatives Fail in South Africa?

South Africa is one of those countries that has a very fertile ground for cooperatives in terms of legislative framework and support, especially since 2004. However, the cooperative sector has remained largely weak and continues to underperform. The Department of Trade and Industry (2012: 10) has pointed out some of the reasons for the failure of cooperatives in South Africa. According to the DTI, the low survival of cooperatives is due to several factors which include the lack of a dedicated agency designed primarily to support co-ops on a focused basis. This results in uncoordinated support that is characterised by poor mentorship, minimal investment dedicated to cooperatives and a sense of neglect among all spheres of government and their respective enterprise development agencies. Other challenges that have been linked to the failure of cooperatives include limitation of current support from existing enterprise development agencies for cooperatives, and lack of recognition of cooperatives as a unique business form (Department of Trade and Industry 2012: 36). There is also avoidance of formalisation by informal self-help groups, accessibility of cooperative registration to local communities, limited promotion and awareness, limited access to critical business infrastructure and limited access to technology. In the remaining part of this chapter, we focus on three crucial aspects of the problems that confront cooperatives in South Africa.

7.4.1 Registration Criteria and Assessment

Some cooperatives on their own have not been fashioned on a genuine basis as they tend to be established for the purpose of accessing cooperative incentives rather than on building a successful cooperative business. This means that people register co-operatives not out of a genuine interest to develop economically and improve their well-being but to access government funding. If one looks at the trend of the sudden increase of cooperatives according to the DTI statistics, it is positively correlated with the establishment of an enabling environment with funding available for emerging cooperatives. That is not to say that there are no cooperatives with a real potential and genuine intention. However, financial incentives (provided to cooperatives in form of grants) incentivise the formation without passionate and dedicated cooperators that can promote the growth of such cooperatives. Such cooperatives are therefore prone to financial mismanagement, lack of member commitment and are likely to collapse once the cooperative no longer access government funds.

It is obvious that the registration criteria and process are not rigorous enough to sift out genuine cooperatives with a potential to succeed. In fact, most of the failures of emerging cooperatives occur during the first year of their registration. According to the DTI report (2009: 38), some of the cooperatives were not active at all which begs the question about what the funding was used for.

7.4.2 Training for Emerging Cooperatives

In their study, Machethe (1990) found that weak management, as a result of lack of training, is one reason for the failure of cooperatives. In addition, the baseline study of cooperatives in South Africa found that

> When asked about training, over 60 % of co-operatives reported that they had not received training in the past two years and if they had received training it was based on general business practices and not sector specific training. A significant number of cooperatives highlighted both general business training as well as sector specific training when asked about their training requirements (DTI 2009).

The study by Pollet (2009) in 11 Africans countries indicates that cooperative-specific training was significant for the success of cooperatives in countries such as Kenya, Tanzania and Uganda. The study found that training was provided either through government programs or by donor agencies. In South Africa, training after registration is largely left to the Small Enterprise Development Agency (SEDA). All other agencies mentioned in the baseline study as providing training to cooperatives do not offer continued cooperative training programmes. Unlike some African countries where donor agencies actively train cooperatives, in South Africa, this is almost nonexistent. The implication of this is that members of most cooperatives are deficient in terms of their understanding of the principles, values, and governance of cooperatives.

7.4.3 Independent Financial Control and External Auditing

A fundamental problem for cooperatives, despite all the support, is a lack of financial discipline and good governance. This is why there is a need for a follow-up of newly established cooperatives. Even though there is financial compliance and report to funding agencies that, in itself, is not enough. One of the requirements for getting financial assistance should be a sound business plan. Firstly, a sector specific analysis of the business plan and support for the development of a business plan is necessary. Secondly, a follow-up check of the cooperative and its projects has to be a condition for awarding grant. To address challenges associated with mismanagement of cooperative finance and other maladministration, the DTI seeks to establish a Co-operative Development Agency and a Tribunal. The development agency and possible satellite sector-specific agencies will have to closely monitor the development of cooperatives and the Tribunal in terms of the enforcement of legislation and dealing with internal matters such as dispute resolution and adjudication. There is also a need for auditing on a regular basis.

7.5 Conclusions and Recommendations

Despite the high failure rate, the cooperative sector in South Africa has gained momentum in recent years. The fact remains, though, that the sector requires a lot of non-financial support. An interesting fact in the baseline study shows that housing cooperatives fared better compared to other forms. The reason for this is that they do not rely, to a large extent, on the government or the market for their existence and development as other cooperatives do. The foregoing underscores the value of the independence of the cooperative sector. In the South African context, there is extensive support of the cooperative which has largely created a dependency syndrome. There should be a clear understanding that the ultimate goal will be for co-operatives to become self-sufficient and operate on business principles.

For cooperatives in South Africa to be successful, it is critical to focus on addressing the three key challenges identified by this review.

The development and growth of cooperatives requires patience. To hasten the process through external funding is counterproductive as these cooperatives will collapse as soon as the external funding dries up. Throwing a lot of money will not bring significant change in the long term. What is needed is a holistic approach that will give cooperatives access to the market and other much needed support. In addition, an organic growth based on communities' realisation of the benefits of cooperation is vital. Such an approach should also emphasise the value of confederation of cooperatives to advocate and champion the cause of cooperatives in South Africa.

A compulsory sector specific training should be given to emerging cooperatives and it should be a condition of any financial support. In addition, there has to be ongoing training that inculcates the principles and values of cooperatives among cooperators. Such training should be geared towards the entrenchment of an understanding of what owing and running cooperatives entail.

References

Arando, S., Freundlich, F., Gago, M., Jones, D. C., and Kato, T. (2010). *Assessing Mondragon: Stability and managed change in the face of globalization*. William Davidson Institute, Working Paper (1003).

Department of Trade and Industry. (2004a). *A co-operative development policy for South Africa*. Pretoria: Department of Trade and Industry.

Department of Trade and Industry. (2004b). *Co-operative development strategy for South Africa*. Pretoria: Department of Trade and Industry.

Department of Trade and Industry. (2009). *The DTI baseline study of co-operatives in South Africa*. Johannesburg.

Department of Trade and Industry. (2010). *Briefing on the Development and Support Programmes for Cooperatives: Empowerment and Enterprise Development Division (EEDD)*. Pretoria: Department of trade and Industry Retrieved 15 November 2012, from http://www.dti.gov.za/parliament/082510_dti_State_of_Co-ops.pdf.

Department of Trade and Industry. (2012). *The DTI integrated strategy on the development and promotion of co-operatives: Promoting an integrated co-operative sector in South Africa 2012–2022*. Pretoria: Department of Trade and Industry, Retrieved 15 November 2012, from http://www.dti.gov.za/economic_empowerment/docs/coops/legis_policy/coop-strategy.pdf.

Department of trade and industry. (2013). *Co-operatives development financial assistance (Incentives)*. Department of trade and industry: South Africa. Viewed 24 March 2014, https://www.thedti.gov.za/financial_assistance/financial_incentive.jsp?id=11&subthemeid=3.

Kanyane, M. H. (2009). Co-operatives as part of social security mainstream for poverty alleviation in selected municipalities. *Journal of Public Administration, 44*(4), 1123–1137.

Machethe, C. L. (1990). Factors contributing to poor performance of agricultural co-operatives in less developed areas. *Agrekon, 29*(4), 305–309.

Nilsson, J. (1999). Co-operative organisational models as reflections of the business environments. *LTA, 4*(99), 449–470.

Nilsson, J., Kihlén, A., & Norell, L. (2009). Are traditional cooperatives an endangered species? About shrinking satisfaction, involvement and trust. *International Food and Agribusiness Management Review 12*(4).

Okem, A. E. (2016). *The Implementation of Cooperative Policy: Perceptions from Cooperatives in the Umgungundlovu District Municipality (KwaZulu-Natal, South Africa)*. Unpublished Doctoral Thesis, Policy and Development Studies, University of KwaZulu-Natal.

Pollet, I. (2009). *Cooperatives in Africa: the age of reconstruction: synthesis of a survey in nine African countries*. ILO.

Republic of South Africa. (2013). Act No. 6 of 2013: Co-operative Amendment Act, 2013. Pretoria: The Presidency. Retrieved 30 January 2014, from http://www.cipc.co.za/Coops_files/Act6of2013CooperativesAmendAct.pdf.

Satgar, V. (2007). *The State of the South African Cooperative Sector*. Retrieved 15 January 2011, from http://www.copac.org.za/files/State%20of%20Coop%20Sector.pdf.

The Companies and Intellectual Property Commission. (2014). *Annual Report 2012/13*. Retrieved 13 January 2015, from http://www.cipc.co.za/files/9513/9989/7347/CIPC_ANNUAL_REPORT_2013.pdf

Van Wyk, C. (2014). East Africa terror hazards tied to poverty: Africa-wide-featured analysis. *Africa Conflict Monthly Monitor* 12–17.

Westoby, P. (2014). Exploring the interface between community development and cooperative development within South Africa—a challenge of theory, practice and policy. *Development in Practice, 24*(7), 827–839.

Chapter 8
The Development and Sustainability of Agricultural Co-operatives at Inanda Township: An Analysis of Inanda Farmers' Association (IFA)

Ndwakhulu Tshishonga and Eleazar Bandyambona

Abstract This chapter explores how the Inanda Farmer's Association contributes towards the development and sustainability of cooperatives at Inanda area in eThekwini Municipality, South Africa. The Association is evaluated in terms of its challenges and strengths in providing institutional support aimed at the development and the sustainability of agricultural cooperatives. The chapter made use of a case study as a research method coupled with interviews of relevant stakeholders and secondary data analysis. The authors in this chapter argue that the failure of community-based cooperatives in dealing with challenges such as poverty, unemployment and inequality is due to lack of understanding of co-op principles and support from cooperatives themselves and other institutions such as government and the private sector. The chapter notes that agricultural cooperatives at Inanda are faced with multiple challenges inter alia, lack of access to land, lack of access to markets and the lack of access to finance. In addition, lack of knowledge about co-operatives, lack of business skills compounded by poor infrastructural facilities were found to be many challenges that impede the progress of cooperatives.

Keywords Cooperative · Agricultural co-operative development · District farmers' association · INK · Poverty · Unemployment · Inequality

8.1 Introduction and Chapter Overview

South Africa has a long and multi-faceted history that shaped the evolution of cooperatives from the racially determined Apartheid cooperatives to the post-apartheid black economic empowerment approach to cooperatives (Satgar and Williams 2011: 202). This protracted history of co-op has revolved primarily in the

N. Tshishonga (✉) · E. Bandyambona
School of Built Environment and Development Studies, University of KwaZulu-Natal, Durban, South Africa
e-mail: Tshishonga@ukzn.ac.za

A.E. Okem (ed.), *Theoretical and Empirical Studies on Cooperatives*,
SpringerBriefs in Geography, DOI 10.1007/978-3-319-34216-0_8

agricultural sector. The South African National Department of Agriculture (2000: 8) states that the cooperative movement in South Africa started in earnest in 1902, just after the end of the Anglo-Boer War. During the War, agriculture in the former Boer Republic of the Orange Free State and Transvaal came to a complete standstill (Van Niekerk 1982). In the Transvaal, the Cape Province, the Orange Free State and Natal, government-backed attempts were made to jump-start the agricultural sector through the cooperative model. However, due to the isolated position of farms, many farmers were skeptical and did not really trust the new development. For this reason, none of the cooperatives could count on the support of their members. Against this backdrop, the beginning of the cooperative movement in South Africa was slow and often fraught with challenges.

Lately, however, a decisive role has been played by cooperatives leading to the establishment of successful farming communities in South Africa. A report from the South African National Department of Agriculture (SANDA 2009) highlights that, when compared with all other means of development, cooperatives have lifted South African agriculture from subsistence farming to a flourishing commercial farming industry. However, Kanyane (2009) argues that there is a need for research into whether or not the cooperative model really offers the best socio-economic development strategy in combating the high rate of unemployment, gross poverty and growing inequality in South Africa.

It is against the above background that this chapter explores the potential role and challenges faced by the Inanda Farmer's Association and the extent to which this Association has contributed towards the development and sustainability of community-based cooperatives at Inanda Township. Section one of the chapter briefly discusses the socio-economic situation in the area. In section two, we present background information about the formation of the Inanda Farmers' Association (IFA). In section three, we present an overview of the research method employed in the study. Section four presents the findings and analysis. This part of the chapter highlights the importance of agriculture in dealing with food insecurity, the creation of employment opportunities and challenges facing the Association. The last section provides the concluding remarks.

8.2 Socio-Economic Situation at Inanda Township

Inanda is located on the Eastern Region of Kwa-Zulu Natal Province, South Africa. It is about 24 km from Durban. Historically, Inanda is one of the oldest areas established in the mid-1800s as a 'reserve' for African people. A sizable Indian population also settled in the area until it was declared a 'released area' for exclusive occupation by African in 1936 (Statistics South Africa 2001 in Department of Provincial and Local Government [DPLG] 2007: 8). Inanda has a population of nearly half a million residents on a 9423 ha and constitutes one of the largest low-income residential areas in South Africa (Khan 2010: 221). The area has a mixture of residential township, a large number of informal settlements and some wards falling within the traditional jurisdiction (eThekwini Report 2004). The DPLG (2007) reports that the population is mostly young with high unemployment, low levels of education, high levels of poverty and crime, inadequate criminal justice capacity, poor traffic and road safety, and land/legal complications. More than a quarter of the population in this node was found to be illiterate (29.7 %) which in turn makes it impossible to break the cycle of poverty as most citizens would struggle to gain employment without basic literate skills and abilities (Buccus 2012: 250). This gloomy situation is further exacerbated by the reality that unemployment at Inanda sits at 66.2 % with 36.4 % of households reported to have no income. In addition, Khan (2010: 221) states that the socio-demographic profile of the location suggests that the population is ravaged by HIV/AIDS pandemic, plagued by increasing rates of unemployment and subject to rapid deterioration in the physical living conditions. These factors are known to negatively impact both livelihood patterns and increasing levels of vulnerability.

Inanda consists of 15 municipal wards, with a large number of informal settlements, limited basic service infrastructure and inadequate recreational/social facilities (Bucuss 2012: 248–249). Similar to other South African townships, Inanda is socio-economically depressed with the 'evil triplets' of poverty, unemployment and income inequality despite national interventions such as the Integrated Sustainable Rural Development Programme (ISRDP) and the Urban Renewal Programme (URP). In their study, Everatt and Smith (2008) found that half of the households (50.5 %) lived in informal or traditional housing. Housing backlogs in Inanda are particularly severe and constitutes almost one-quarter of the backlog in eThekwini Municipality (eThekwini Report 2011). Inanda is the mirror of socio-economic predicaments faced by eThekwini municipality.

Although there has been significant progress in some areas, there is still a huge need to address challenges such as poor access to basic household and community services, high levels of poverty, crime and risk, and illiteracy, low levels of skills development, lack of adequate healthcare facilities and the unsustainability of development practices (DPLG 2007 and eThekwini Report 2011).

8.3 Background of Inanda Farmers' Association (IFA)

The IFA is a secondary cooperative apex body that comprises of five small associations and has been in existence since 1997. The association was formed with the primary purpose of improving the functioning of agricultural cooperatives within the Inanda area in dealing with the problem of food insecurity (Ismail 2014). The Inanda Farmers' Association is not only charged with the responsibility of promoting gardening through agricultural cooperatives, but also serve as a mouthpiece and a link between its members and relevant government departments at the eThekwini Municipal level and KwaZulu-Natal provincial level. According to eThekwini Innovations (2007: 2–3), the Association has the following objectives:

- To increase and formulate the business base in rural and semi-rural areas, thus resulting in job retention and growth
- To accelerate and broaden the base of agricultural skills development, which will lead to greater entrepreneurial activity
- To provide new channels of both business-to-business and business to government linkages between age agricultural areas and business centres
- To provide an opportunity for increased food security, at a municipal scale, but also in time at a national scale
- To provide a platform for increased research and commercialisation of biotechnology opportunities, for example ethanol and bio-diesel
- To create new investments in previously disadvantaged areas
- To give a platform for synchronised co-operatives, shared resources and a common strategy across the three spheres of government, and
- To increase the city's economic activity within the agricultural sector
- To commercialise community gardens, thereby increasing the disposable income of those who tend them.

These objectives can only be achieved through co-ops mobilisation and strategically repositioning themselves to sustain an association like the IFA. Interesting outcomes of the case of the Inanda area is that while the Inanda, Ntuzuma and KwaMashu (INK) Agricultural and Multipurpose Secondary Cooperatives has been in existence for some time, very little has been achieved. Hence, the unification of the five associations from the Inanda Farmers' Association (IFA) in the eThekwini Municipality as a strategy to improve their households' socio-economic situation (Bandyambona 2013: 7). Currently, the Association is managed by 11 board members selected from the member co-ops which has approximately 200 co-op members including both individual farmers and co-ops (Ismail 2014). Individual or primary cooperatives interested in joining the Association are required to pay R150-00 for their membership and the money is often used to keep up the daily running of the Association including transportation to attend meetings (IFA Constitution, 2). The progress and achievements of this secondary cooperative, since its inception in 1997, has been very slow. Some of the achievements include registration as a legal entity with CIPC, securing permission to utilise three vacant offices in Newtown A, and permission to operate as a flea market in the KwaMashu E section.

8.4 Methodology

Qualitative research method was employed in collecting data for this study. The triangulation of research techniques such as document analysis, individual interviews and observation were implemented. In this regard, Tsheko (2007: 6) argues that the use of multiple sources of research data such as semi-structured interviews and document analysis do not only provide an essential descriptive data, but also make it possible for the triangulation of data analysis and documenting purposes. Semi-structured interviews with individuals were conducted with key stakeholders who are board members of the Inanda farmers' Association (IFA). Three sets of visitations were made to the board members at Inanda and interviews were conducted in their natural setting. 7 Board members including 2 cooperative members were the target sample and were selected purposively. In total, there were 9 participants in the study. Each interview took approximately 45 min. The data collected for the study were analysed thematically. Key findings are presented according to emerging themes in the following sub-sections. Direct excerpts from interviews are italicised.

8.4.1 The Importance of Agriculture for Low-Income Households

Subsistence agriculture is an important sector in reducing extreme poverty and food insecurity in low-income areas such as Inanda. The Reconstruction and development Plan (1994: 102) states that a vibrant and expanded agricultural sector is a critical component of rural development and land reform programme. Unfortunately, through the Land Act of 1913, only 7 % of the country's land was reserved for 'natives' which eventually expanded to 13 % through the 1936 Native Trust (Atuahene 2014: 8). This has limited the participation of black South Africans in agricultural sector. Atuahene (2014) reports that agriculture does not only contribute 5 % of the GDP but also add over 10 % of employment. The triple challenges of poverty, unemployment and inequality made the government and other interested stakeholders prioritise food production as one of the mechanisms to deal with food insecurity. A young emerging farmer highlighted the desperate situation in which poor people found themselves without food. He went on to say that:

> Food insecurity has multiple effects on the poor. Without adequate and nutritious food, the poor and destitute do not only suffer from health related diseases such malnutrition, but also lack of it weakens their immune system to an extent that they are unable to be productive.

In line with the foregoing, the chairperson of the IFA noted that *agriculture plays a pivotal role more particularly in feeding people with fresh vegetables, cutting down poverty and creating job opportunities for the urban poor.* Judging from the foregoing, it is not surprising that the Province of KwaZulu-Natal, through the Department of Economic Development in partnership with the Department of Agriculture, has initiated a programme called 'One Home-One Garden' to encourage people to plant their own food.

According to the Association's secretary, the association specialises in crops such as potatoes, cabbages, carrots, green beans, spinach, and onions. Other Board members of the Association stressed that farming in crops is important in communities threatened by food insecurity coupled with poverty, unemployment, income inequality as well as the scourge of HIV/AIDS. One farmer asserted that:

> All these problems are depressing but through self-initiated enterprises such as the cooperatives, people start to think differently and to believe in themselves. Small-scale farming does not only pave way to the eradication of poverty and unemployment, but also enable farmers and general communities to get food and some money.

8.4.2 Opportunities of Primary Co-ops Through Inanda Farmers' Association

Since inception, the IFA has created opportunities which, when exploited, could make cooperatives strong, effective, efficient and proactive movement capable of providing most needed goods and services. The availability of land at Inanda is the primary resource and without it, it will be impossible for people to pursue agricultural activities. Government (the Department of Agriculture in particular) has been helpful in the provision of agriculture-related facilities such as tractors, seeds, and technical support to various cooperatives in the area. In addition, the Department offers grants to both individual and registered co-operatives.

At Inanda, job opportunities are very scarce and local communities are pushed to organise themselves into small income generating projects as part of their livelihood strategies. As alluded by the chairperson of IFA, agriculture is one of the sectors which has the potential of creating jobs for local people. It could also render needed goods and services to respective communities. The members of the Association were appreciative of the formation of agricultural cooperatives as reflected in the interview excerpts below:

> Although other people in the community consider small-scale farming as a lower-class occupation, tilling the land reconnect us to earth and help us to earn a living thereby creating employment opportunities.
>
> Working the land does not only teach skills on how to deal with soil and the environment and other agricultural related facilities and equipment, but also restore our human dignity and a sense of active citizenship.

During the interview, the Treasurer of the Association revealed that 124 people are employed in the 17 gardens supported by IFA. He also reported that they used approximately 15–20 ha of land. Importantly, co-ops do not only play a unifying role of different ethnic groups in their localities (Fairbairn 2004: 29), they also provide employment opportunities thus ensuring that agriculture has widespread and powerful social and economic effects on people's lives. One of the co-op members highlighted the prevalent socio-economic challenges motivated the Association to promote self-help and improvement of farmers who in turn would provide agricultural goods and services. Members of the Inanda Farmer Association, each owns a piece of land where s/he grows different crops according to the market demand.

Improvement of the household socio-economic situation was another opportunity provided to members of the IFA. Most of the interviewees noted that the crops they produce as co-ops help them to improve their socio-economic status. Money generated from the sale of produce is used to purchase household goods. It was also noticed that children of most members of the cooperatives were attending school. This was a good sign for the future of the households.

Linking other co-ops around Inanda to agriculture market was also identified as an opportunity provided to primary co-ops by the IFA. Since the Inanda Farmers' Association is a formal organisation, it links with other organisations/associations both within and outside Inanda to exchange and share ideas. The chairperson of the association reported that there is high demands for co-ops to produce and it became clear that co-ops needed to organise in order to meet the demands. Apart from selling their produce within their respective communities, the Association has managed to secure orders from retail shops such as Spar, Shoprite and Pick'n Pay.[1]

8.4.3 Challenges Facing District Farmer Association

Analysis of field interviews identified a number of challenges facing the IFA. These challenges are presented below.

8.4.3.1 Shortage and Limited Optimal Use of Land

Land is a big issue at Inanda as the government does not own much of land, individuals have small portions of land which they use for various purposes such as the construction of houses and businesses or small agricultural farming. Previously, these small farmers operated as associations but without noticeable impact especially on embarking on effective and efficient subsistence farming. During the interview, the chairperson of the association noted that:

> The shortage of land for members of the Association and INK community, in general, should be measured on how effective and efficient individual farmers and cooperatives are making use of the available land at their disposal. The size of land owned for agricultural use varies; some individuals have between 1, 2 and 5 half hectors while co-ops own between 2, 3 to 5 hectors of land.

Most of the cooperators interviewed highlighted that only a few farmers were in dire need of extra land, the rest of cooperators reported that the challenge rest on the optimal utilisation of the available land. Little has been done by the local government to support these associations to maximise land use. Given its status as one of five area based management areas in the city, Inanda is part of Inanda, Ntuzuma and KwaMashu (INK) and its proximity to uMhlanga and the northern

[1]All are major retail chain stores in South Africa.

development corridor made it to become more integrated into greater Durban and its metros in the northern spatial plan. The demand for land in this area, especially for business purposes, makes it very difficult for people to venture into expanded agricultural activities. Against this backdrop, maximizing the productive of land holding if critical to the success of small-scale farming.

8.4.3.2 Financial Support and Abuse of Government Grant

Economic activity that would allow for large-scale employment generation is non-existent at Inanda, while the majority of consumer spending leaks out of the Inanda area. For individual farmers, limitation to finance resources poses a huge challenge hence the majority of the households are not formally employed and depend on government grants. For most of the poor and unemployed, the formation of agricultural co-ops was a relief but lack of proper planning, as well as generic start-up financial support, resulted in poorly managed cooperatives with little financial outcome for members. One member of the Association revealed that government assisted them with financial support in the form of grants. Unfortunately, such assistance is being abused by cooperators who spend the money without saving or investing for future use. It has been reported that some members desert farming and seek employment elsewhere once the grant is exhausted. Such members eventually come back when there is the promise of a further grant from the government. Satgar and Williams (2011: 207) applaud the assistance of a grant or loan-based financing model, but also warned that it may entrench dependency on the state. It is also important to highlight that due to the lack of proper organisational system and good planning; it is difficult for the Inanda community to access finance as finance institutions require formal organisation structures.

8.4.3.3 Lack of Market and Networking Skills

Since cooperatives are based within their communities, local people and communities provide co-ops the primary market. In the past, the Association managed to secure formal market which entailed supplying retail outlets such as Pick'n Pay, Checkers, Spar, and Woolworth. The main challenges in the agricultural sector is to compete with bigger businesses with more expertise, capital, and means of mass production. The following comments shed light on the imperative for creativity, innovation and strategic repositioning of co-ops to sustain their entities as part of their competitive advantage.

> This sector does not entertain people who are lazy hence most of co-op members sit and expect good harvest without sweating for it.
>
> Farming demands hard labours and people who are willing to go an extra mile. This includes ploughing on time; ensure plants are taken care of in order to guarantee good harvest.

> Instead of waiting for an order, farmers have to take into themselves to use informal marketing strategies such as going house to house selling, sell to hawkers and in the local taxi ranks, etc.

The reality is that there is demand and co-ops are failing to meet these demands. From the interviews, it emerged that in cases where co-ops have secured a market, their produce is not usually supplied on time and the quality of produce is often considered sub-standard. One of the members stated that in order to deal with marketing the co-ops' produce, it is imperative for the Association to create a market around the area for the local community to be able to support the project in buying the produce. One of the executive members of the Association states that:

> The Association is faced with multiple challenges. Central to these of the challenges is the failure by the Association to secure sustainable markets to sell the produce. As an Association, we do not have marketing skills, attractive packaging of produce and the failure to master technology such as computers and the internet which contribute towards undermining our capacity to market our produce.

Due to the high level of unemployment in the Inanda area, the crime rate is high. The level of security is very poor and produce ready for harvest is often stolen. Agricultural equipment such as machinery, and seedlings and manure need 24 h security. This adds additional operating costs to the activities of the cooperatives.

8.4.3.4 Transportation of Products to the Market

Transport of the farmers' produce is another challenge. Considering that most of the crops are perishable, there is a need to transport them to the market timeously. Good roads are needed to aspire and attract both locals and bigger supermarkets to visit the farmers and support the cooperatives. For the survival of the Association and its individual member cooperatives, road infrastructure is an imperative while the availability of transport is paramount for moving produce to the market. The following excerpt captures a member's view on challenges associated with transport:

> We find it very difficult to transport our product because we do not have transport. At times, we hire other people's cars who in most cases charge us exorbitant fares because they know that we do not have any other alternative. Without our own transport, we are often

To address the challenge of transport, it was noted that some members have improvised by taking their produce to the taxi rank and local schools in order to sell. Considering that most of the member cooperatives produce perishable goods such as cabbages, potatoes, carrots, spinach and tomatoes, inability to timeously move these to the market would result in significant loss for the farmers.

8.4.3.5 Lack of Understanding of Co-op Values and Principles

One of the challenges facing members of the association is limited understanding of the principles of a cooperative and how co-ops differ from other enterprises.

However, it was noted that there were ongoing attempts to bridge this knowledge gap as members undertake training to improve their knowledge.

8.4.4 Farming Equipment and Tractors

For the Association to execute its work effectively, they need equipment and ploughing tools such as tractors, seeds, irrigation facilities, etc. During the interview, it was revealed that the eThekwini Municipality tried to help the Association with farming equipment such as tractors but due to the high demand of the materials, cooperators were made to wait for longer periods in order to be assisted. One of the cooperators highlighted his frustration in the following excerpt:

> Despite that we are able to access the municipality owned tractors, the challenge is that more often the tractors is not released in time for the farmers to take advantage of such resource.

In response to this dire situation, the members were compelled to go out and rent tractors which affected their profit margins. This is expressed in the excerpt below:

> In most cases municipality would release the tractor for two days while the farmers need the tractors for a week. This situation then pushes us to hire outside tractors for them to do the work.

Considering that farming is seasonal, the Association's failure to plough on time has serious implications for productivity. Produce such as potatoes, onions, cabbages, carrots are seasonal and demand a tight programme pertaining to preparation for planting, weeding and harvesting.

8.4.5 Quality Seeds and Ploughing Methods

The eThekwini Municipality, in collaboration with the provincial Department of Agriculture, ensures that people and organisations in the farming sector are taught effective ploughing method and necessary facilities are provided for quality produce. Unfortunately, the farmers highlighted that they were currently using traditional farming methods including the irrigation system which enables them to meet the demands from customers. Some interviewees wanted more support from either the government or donors especially financial support to purchase seedlings. They reported that they were unable to keep seeds from the current harvest as leftover seeds were not good for replanting hence they needed money to buy more for the next season. One of the solutions suggested by the Vice-Chairperson of the Association to this problem was for the Association to start keeping their own seeds and learn to preserve seeds with assistance from the Department of Agriculture.

8.5 Discussion of Research Findings

Despite the developmental interventions such as the Urban Renewal Programme Area-based Management programme, Inanda is still categorised as one of the most deprived and low-income areas in South Africa in terms of basic services and food insecurity (DPLG 2007). Inanda is plagued by multiple challenges hence growing food could be instrumental in mitigating malnutrition and hunger, supplement the diet and provide employment for the locals. Cooperatives that produce a variety of alternative and locally desired items could be a useful tool and agents of change in their efforts to enhance rural communities and their economic well-being (Wilkinson 1991; Luloff and Bridger 2003). Most of the cooperators agreed that cultivating the land at this scale gives them economic power as they could sell cultivated produce for household and public consumption. They added that vegetables could be used to sustain their livelihoods and contribute towards reducing hunger in the area. In this case, well organised, financial sustaining and properly managed cooperatives could be a force where employment opportunities are created; reliable income generated and increased trade. Further, cooperatives serve to strengthen community support functions.

Despite the real and potential benefits of the IFA, it is constrained by a number of challenges. In the case of the Inanda Farmers' Association (IFA), the shortage of financial resources coupled with the lack of land, farm input, and access to market, transport and knowledge on cooperatives have made it difficult for the Association to function. These challenges are further aggravated by the dominant spirit of individualism as opposed to cooperation and self-help within the Inanda area. Division and conflict of ownership were dominant among the members. This has made it difficult for board members to work cooperatively with members as recommended by Ginder and Deiter (1989: 318).

KwaZulu-Natal has the highest number of cooperatives but the majority of them collapse without yielding fruits. Lack of bottom-up cooperative development centred on member education has resulted in dismal outcomes evidenced in 88 % failure rate (Satgar and Williams 2011: 205). Some of the reasons attributed to the failure of co-ops were that people were lured into forming co-ops because the KZN province promises financial support (Satgar and Williams 2011). They further noted that

> Instant registration of co-operatives without proper planning, generic start-up financing and supply-side attempts at co-operative education have not led to genuine co-operative consciousness and understanding. Instead, co-operatives have been treated as just another business form and understood in quantitative terms as part of out-based management within the state bureaucracy (Satgar and Williams 2011: 205).

The problem of financial incentive as a barrier to the success of cooperatives was noted in interviews conducted with members of the IFA. Members cited instances where some individuals join cooperatives only when the government has provided grants to cooperatives.

Added to the foregoing is limited understanding of cooperatives and how they function. At the movement, cooperatives rely largely on the government for training on the principles, values, and governance of cooperatives. Such an approach furthers the dependence of cooperatives on the government for their survival. For that reason, the apex body like the IFA needs to take a proactive step towards member education. Such self-reliance should also extend to other areas of the operations of cooperatives. For instance, as much as the Association accepts the help from the Department of Agriculture, it is imperative for this apex body to organise itself so that it could be self-reliant in small items such as seeds. For the co-ops to compete and sustain their enterprises, it is imperative that the Association organise a series of workshops where farmers could be taught innovative methods of farming including the use of chemicals and pesticides.

8.6 Concluding Remarks

This chapter focused on the roles of Inanda Farmers' Association in developing and sustaining the cooperative sector. The findings show that agricultural cooperatives at Inanda have the potential to impact positively on the socio-economic status of rural people and other low-income areas as long as cooperative values and principles are respected by all cooperators. Cooperatives at Inanda contribute, to a limited extent, towards job creation, income generation as well as better living conditions of Inanda community. The findings of this chapter demonstrated that for cooperatives in Inanda to be sustainable, cooperatives themselves, including their apex body such as Inanda Farmers' Association, should take the primary responsibility for the cooperative. Government and its Cooperative Unit should assume the secondary role especially in consolidating cooperative policy and educating cooperatives on the essence of the cooperative model in reducing poverty and job creation.

The professional monitoring of agricultural activities within the Association is imperative. At Inanda, monitoring and evaluation of produce is unfortunately poor and weak. For the Association to be sustainable, professional services regarding monitoring and supervision could be instrumental in maintaining quality and meeting market demands. Although the Department of Agriculture provides technical support to individual co-ops, such support should allow space for the Association to organise itself in order to avoid dependency. Thus, support by government should either be in the form of facilities, grants or training and education programmes and should not undermine co-op principles of autonomy and independence and economic control including values such as self-help and solidarity.

The chapter found that cooperatives at Inanda are confronted by a number of challenges including financial and infrastructural deficiencies which undermine their ability to contribute to poverty reduction and employment creation. Challenges such as shortage of land, financial constraints, coupled with poor understanding of what and how cooperatives function, as well as their values and principles, are the stumbling blocks undermining the potential of IFA in growing a vibrant and capable cooperative movement. In order for the cooperatives at Inanda to reach their objectives of poverty alleviation and job creation, there is a need for targeted intervention such as the creation of viable markets. eThekwini municipality could recommit itself to consolidating its preferential tendering system whereby cooperative enterprises could be prioritised. In return, this commitment could be instrumental in providing employment and create sustainable incomes through the consistent production of goods and services.

References

African National Congress (ANC). (1994). *The reconstruction and development programme: A policy framework*. Johannesburg: Umanyano Publications.

Atuahene, B. (2014). *We want what's ours: Learning from South Africa's land restitution programme*. Oxford: Oxford University Press.

Bandyambona, I. (2013). *Community-based co-operatives in Inanda, Ntuzuma and Kwamashu (INK) community (eThekwini municipality) as an alternative form of economic development: Lessons from the Kenyan co-operatives models*. A master dissertation submitted in the School of Built Environment and Development Studies. Durban: University of KwaZulu-Natal.

Buccus, I. (2012). Community life and securing participation beyond elections'. In S. Booysen (Ed.), *Local elections in South Africa; parties, people, politics*. Bloemfontein: Sun Press.

Department of Provincial and Local Government and Business Trust. (2007). *Inanda, Ntuzuma, KwaMashu (INK) Nodal Economic Development Profile*. KwaZulu-Natal.

Department of Agriculture, Forestry and Fisheries. (2011). *Annual Report on the Status of Agricultural Co-operatives*. Pretoria: Department of Agriculture, Forestry and Fisheries.

eThekwini Municipality. (2004). Making City Strategy Come Alive: Experiences from eThekwini Municipality, Durban, South Africa 2000–2004. eThekwini Municipality.

eThekwini Municipality. (2007). *eThekwini Agricultural, Arts 7 Craft Expo 2006*. eThekwini Municipality.

eThekwini Municipality. (2011). *eThekwini Co-operative Development Strategy*. Durban: eThekwini Municipality Business Support and Markets Unit.

Everatt, D., & Smith, M. J. (2008). *Building livelihoods in Inanda: A report submitted to the National Department of Social Department*. Johannesburg.

Fairbairn, B. (2004). History of Cooperatives. In C. D. Merrett & N. Walzer (Eds.), *Cooperatives and local development: theory and applications for the 21st century*. London: M.E. Sharpe, Inc.

Ginder, R. G., & Deiter, R. E. (1989). Managerial skills, functions and participants. In D. Cobia (Ed.), *Co-operatives in agriculture*. Englewood Cliffs: Prentice-Hall.

Ismail, M. (2014). Chairperson of the Indanda Farmers' Association, Interview on the 19 August 2014.

Kanyane, M. H. (2009). Co-operatives as part of social security mainstream for poverty alleviation in selected municipalities. *Journal of Public Administration, 44*(4), 1124–1137.

Khan, S. (2010). Challenges facing poverty relief projects in the urban renewal programme of Inanda in the eThekwini Municipality. In M. S. Mapadimeng & S. Khn (Eds.), *Contemporarty Social Issues in Africa: Cass in Gaborone, Kampala and Durban*. Pretoria: Africa Institute of South Africa.

Luloff, A. E. & Bridger, J. (2003). Community agency and local development. In D. Brown & L. Swanson (Eds.), *Challenges for rural America in the twenty-first century* (pp. 203–213). Pennsylvania: Pennsylvania State University Press.

Satgar, V., & Williams, M. (2011). The worker cooperative alternative in South Africa. In J. Daniel, P. Naidoo, D. Pillay, & R. Southall (Eds.), *New South African review 2: New paths, old compromises*. Johannesburg: Wits University Press.
Tsheko, G. N. (2007). *Qualitative research report on orphans and vulnerable children in Palapye, Botswana*. Cape Town: Human Sciences Research Council Press.
Van Niekerk, J. A. S. (1982). *A review of the development of agricultural co-operation in the republic of South Africa*. Silverton: Promedia Publications.
Wilkinson, K. P. (1991). *The community in rural America*. New York: Greenwood Press.

Chapter 9
The South African Cooperative Sector: Which Way Forward?

Andrew Emmanuel Okem and Ndwakhulu Tshishonga

Abstract The South African cooperative sector continues to be confronted by challenges which have undermined their success despite the concerted support of the sector in the past two decades. To further the success of cooperatives, it is critical to explore the theoretical and empirical understanding of cooperatives in order to glean lessons that could be applied to the South African context. In this chapter, we argue that for cooperatives in South Africa to truly become sustainable, they must work towards the radical application of the principles and values of the cooperative movement. This entails eliminating the practice of dependence on the state and ensuring that cooperatives are linked to addressing socio-economic issues for which they are established. Additionally, there is an urgent need for increased understanding of the rationale for the establishment of cooperatives through education and training of cooperators. The chapter further argues that there is a need to allow the organic growth of cooperatives instead of the government facilitating a false growth through unsustainable financial and non-financial incentives.

Keywords Cooperative · Dependency · Self-help · State-led · Successful cooperatives

9.1 Introduction and Chapter Overview

This book sets out to assess theoretical and empirical perspectives on the cooperative movement in order to establish possible lessons for the cooperative movement in South Africa. Chapters in this volume explored theoretical debates and empirical

A.E. Okem (✉) · N. Tshishonga
School of Built Environment and Development Studies,
University of KwaZulu-Natal, Durban, South Africa
e-mail: okem@ukzn.ac.za

N. Tshishonga
e-mail: Tshishonga@ukzn.ac.za

A.E. Okem (ed.), *Theoretical and Empirical Studies on Cooperatives*,
SpringerBriefs in Geography, DOI 10.1007/978-3-319-34216-0_9

studies on factors that contribute to the success/failure of cooperatives both in and outside of South Africa. This chapter presents the key lessons that could be applied to the South African context. The chapter is divided into two sections. In the first section, we provide a summary of emerging findings from the chapters presented in the book. The second section apply these findings to the South African context. The chapter notes that recommendations presented in the book could go a long way in addressing multiple challenges facing the cooperative movement in South Africa.

9.2 Theoretical and Empirical Perspectives on the Cooperative Movement

The cooperative movement has a long history as a grassroots movement that has played a critical role in addressing people's poor socio-economic conditions around the world. The emergence of modern cooperatives in the UK at a time when the imperatives of the industrial revolution left many destitute is a testament to the power of collective action in addressing societal needs. Although the cooperative movement has and continues to be confronted by a number of challenges, the successes it has recorded over the years in lifting millions out of poverty through the provision of livelihood options and building social capital all demonstrate the relevance of the sector to present day global challenges of poverty and unemployment. Although the ideals of creating a society where the values and the needs of individuals trump that of profit maximisation continues to be utopian, the success of cooperatives have demonstrated that it is a viable economic alternative particularly in contexts where the market has failed to provide decent livelihood options. The review of the global cooperative movement demonstrated that the sector is not limited to small-scale business activities of poor vulnerable communities. Cooperatives have and continue to engage in large-scale businesses and are contributing to the economies of countries where they operate.

The historicity of cooperatives in Africa showed that although modern operatives traced their origin to the industrial age in the UK, traditional forms of cooperative pre-existed this modern forms in many societies. These traditional forms of cooperatives are organic bottom-up associations that share many of the values that define the cooperative movement. However, these traditional forms are not formally recognised and have not been integrated into the formal economy despite their valuable contributions to members' social and economic conditions. The implication of this is that such informal organisations do not benefit from policies and support that could further enhance their success. Additionally, the formal cooperative sector are not able to benefit from the values that have been critical to the success of these informal organisations due to the lack of intersection between the two sectors.

Modern cooperatives that were introduced in colonial Africa have had varying success in terms of their impact on socio-economic conditions of people. The historical background on the evolution of cooperatives demonstrates that although the modern form of cooperative was introduced by the colonial government, postcolonial African governments adopted them as a means of addressing multiple challenges in many of the then newly independent African states. While such approach was instrumental in the growth of the cooperative sector, it did not translate into the success of cooperatives. The growth of cooperatives recorded in post-independence African countries was a false one since the focus was on the number and size of the sector rather than on the extent to which cooperatives are able to operate independently of government, empower local communities and contribute to socio-economic development. In addition, the focus of many African countries was to use cooperatives as a tool for achieving political ends rather than allowing them to emerge organically and operate according to the principles and values of the cooperative movement. The Kenyan and Cameroonian scenarios both demonstrate that over-reliance on the government for the success of cooperatives leads to the inability of cooperatives to be a successful enterprise. Recognising the linkage between state dependency and the failure of the cooperative sector precipitated the transition to the removal of the dependence of cooperatives on the state. This approach has been characterised with varying outcomes such as the revitalisation of the cooperative sector in Kenya where cooperatives now compete in financial and non-financial sector with private businesses to the complete failure of cooperatives in Senegal.

Despite the negative outcomes of the liberal economic approach to the cooperative sector, the approach has undoubtedly contributed to the renaissance of cooperatives across the continent in recent years. For instance, the independence of the sector led to the rediscovery of the cooperative movement as a member-owned and member controlled sector. In addition, there is an increasing recognition that cooperatives are organic associations of persons whose operations and survival are dependent on members' economic participation. This has ensured that cooperatives in most African countries are now aligned (to varying degrees) to the internationally recognised principles and values of the cooperative movement. Increasingly, cooperatives are beginning to operate in a space that is conducive to their growth rather than being captured by the state. In the new space that cooperatives find themselves, they are beginning to make alliance among themselves as well as in the private sector for successful business operations.

Cooperatives, because of their focus on socio-economic development, have been recognised as a valuable tool in post-conflict reconstruction. It is noted that since conflict often results in the destruction of people's livelihoods, improving economic conditions is a critical component of post-conflict reconstruction efforts. The role of cooperatives in this context is not only limited to economic growth but also the building of social capital. By creating the context where people work collectively to meet their common needs, cooperatives create opportunities for dialogues, trust building, and peaceful co-existence.

9.3 Drawing Lessons for the South African Cooperative Sector

The chapters presented in this book explored cooperatives from the theoretical and empirical perspectives with the goal of outlining possible indicators that could be contextualised in South Africa for the success of the country's cooperative movement. While the success indicators presented in this summary does not pretend to be a panacea to the challenges facing the country's cooperative sector, it is hoped that they will contribute to ensuring an increase in the understanding and consolidation of cooperatives in the country.

9.3.1 Cooperatives Claiming Their Social Movement Status

South Africa could learn from the Kenyan and Cameroonian experiences especially given the reality that there is little development and growth through the current state-led approach to the development of cooperatives. The Kenyan and Cameroonian scenarios both demonstrated that government intervention and interference with the cooperative movements created a warped notion of cooperatives which led to the underperformance of the sector. In both Kenya and Cameroon, the state-led approach undermined the ability of cooperatives to adjust to the demands of an open competitive market economy. Rather than growing into an active sector that promotes members' socio-economic interests, the state-led approach in both countries facilitated a growth in the number of cooperatives without a corresponding increase in their contributions to the economy. The removal of state support and the decline of cooperatives in both countries underscores the weakness of the state-led approach.

In the South African context, the resurgence of cooperatives in recent years has been mainly characterised by excessive involvement of the state in the cooperative sector. The high failure of cooperatives, despite ongoing government support, indicates that the South African cooperative sector will continue to remain on the fringes of the economy if the current development trajectory remains unchanged. Since the early 1990s, most newly established cooperatives have collapsed and have suffered multiple blows with the lack of financial support and absence of a market. In addition, most cooperatives are fraught with leadership crises. Like the Kenyan case, South African cooperatives could benefit enormously from maintaining their social movement status. This would enable them to forge a beneficial partnership with either the state or the private sector. As a social movement, co-ops will not only maintain their independence but will be at liberty to challenge prevailing socio-economic and political injustices and use their alliance to influence government policy direction. For the sustenance of cooperatives, it is imperative that cooperatives constantly fight for their autonomy and self-management. In this context, the ICA advocates that the cooperative movement plays a crucial role in influencing the relationship between the co-ops and the state through demanding for more freedom and independence from the state.

9.3.2 Radical Application of Co-ops' Principles and Values

In the case of Kenya, the state was active in guiding and directing the formation of cooperatives in both the colonial and post-colonial era. This state-led approach led to the subordination of cooperatives to the whims and caprices of the state. It was only when the state withdrew its active role in the running of cooperatives that Kenya saw the cooperatives thriving in mobilising people socially and economically through the advancement of its principles and values such as voluntary and open membership, democratic member control, member economic participation, education, training and information, co-operation among co-operatives and finally its concern for community.

In post-Apartheid South Africa, most cooperatives grew in response to the financial incentives from the government. In addition, cooperatives largely rely on government institutions for member education, linkage to market opportunities as well as other inputs required for their operation. The radical application of co-op principles and values could result in the South African cooperative sector becoming proactive in addressing the challenges they face rather than relying on the state and non-governmental organisations.

9.3.3 Linking Social to Economic Development

In the case of South Africa, Cameron (1996: 324) argue that the development of cooperatives had to deal with the complex legacy of social, economic and political exclusion. It is therefore not surprising that the mushrooming of co-ops took place without any orientation on what the co-op model is all about. Viewed from the generic term of community enterprise, cooperatives tend to be a practical way of mobilising local action and resources in the local economy for the benefit of people. In essence, the application of cooperatives values and principles could place them at an advantage to regenerate local community, deliver local services while at the same time tackling social problems.

Cooperatives in a socially fragmented society like South Africa can play a pivotal role not only as an economic vehicle for the people to generate income but can be avenue where members can share knowledge, discuss social ills that can bring about trust and peace building. Cooperatives can be a vehicle for addressing social ills such as racism, xenophobia, and crime that have come to be associated with South Africa. The example from Rwanda demonstrated the valuable roles of cooperatives in reconciliation and building of social capital. The violence of the Apartheid era as well as those that characterised the years of transition have left a deep sense of distrust which could be addressed through the radical application of cooperatives as a tool for social transformation. To achieve this end, it is of utmost importance that there is a sound understanding of the meaning, aims and objectives of cooperatives and that these be applied in the implementation of this approach.

9.3.4 Utilising the Strength of Network and Supply Chain

In terms of the production chain, social capital through bonding and bridging could afford cooperators to network and forge social as well as economic networks in order to broaden and unlock their creativity and innovations. Linking social and economic development is also underpinned by an integrated grounded development intervention where development does not only prioritise economic growth but also look at other dimensions such as social, political and cultural aspects of development. Kenyan cooperatives have experienced various challenges and have transitioned from being state-led to being member-owned and member-controlled. This has enabled them to build strong linkages among cooperatives. The linkage has played a critical role in the success of cooperatives as they eliminate transaction costs and ensure that member cooperatives are competitive in the market. This approach was made possible through the creation of an enabling environment that allows cooperatives to flourish on their own.

Although the South African scenario shows that the government is cognisant of the need to create an enabling environment for cooperatives, its actions show that it has and continues to be deeply in charge of the cooperative movement. This is evident in the failed attempt to establish the South African National Cooperative Alliance as well as the ongoing financial support of cooperatives despite available evidence which demonstrates that financial incentives are not advancing the success of cooperatives. In the presence of extended support, the incentive to form strategic network for success is significantly weakened among cooperatives. South Africa could learn from Kenya by restricting its activities mainly to that of facilitatory role.

9.3.5 Uprooting Dependency Through Self-help

History shows that cooperatives in Kenya were not always successful because they were controlled by the government which administered and supervised cooperatives. The review of the history of cooperatives in Kenya showed that state control of the sector stifled growth and innovation (Ogina in Pivot 2009: 4). A similar scenario was pointed out in the review of cooperatives in Cameroon. In the Kenyan scenario, the withdrawal of state control made cooperatives autonomous, self-reliant and commercially viable organisations (Ogina in Pivot 2009). The Kenyan government also created an enabling environment through the Co-operative Societies Act of 1997 (amended in 2004). In addition, the government gave cooperatives tax relief. This approach could be adapted by the South African government in order to facilitate the growth of the cooperative sector.

Although cooperatives in Kenya were first initiated by the colonial government to market crops produced by smallholder farmers , they took off to become the most successful employers for many Kenyans. Muthuma (2011) states that cooperatives in Kenya operate on a commercial basis and are economically strong. The success and strength of the Kenyan cooperative sector has been linked to the understanding of cooperatives as self-help businesses rather than as means of accessing government funds.

In the South African context, government's approach seems to be hinged on political expediency rather than on seeing the sector as a viable economic sector. Such a view can be linked to the ongoing support despite the fact that the sector has largely underperformed despite extensive support. South Africa can learn from the cooperative history of Kenya that cooperatives should be run as businesses that have a strong social objective. This approach will contribute to eliminating the current scenario where many cooperatives are mismanaged by members who lack business principles and a sound understanding of cooperatives as self-help businesses.

9.3.6 Advancing Coops Through Training and Education

Since the introduction of the Co-operative Act of 2005, the South African cooperative sector has been characterised by limited training and education of members of cooperatives. This resulted in confusion about the meaning of co-operatives and its social and economic advantages. Although there are existing training and education support from the government, these initiatives have been hindered by a lack of skills and competent personnel. In addition, the services are fragmented and only available to a certain section of the population. Often, these services are not available to rural communities. It is therefore recommended that co-operatives should organise themselves into a formidable force and establish a co-operative college aimed at capacitating co-operatives.

A co-operative college could mitigate the trend of collapsing and defunct/mismanaged cooperatives in South Africa. In the Kenyan scenario, the cooperative movement is not only anchored on the sector being autonomous and member-controlled; the sector also equips members with training and skills through the Co-operative University College of Kenya (Muthuma 2011). The college has been offering quality cooperative training about cooperatives with a specific focus on cooperative management, law, accountancy and cooperative banking at certificate, diploma and degree levels. South Africa is in dire need of a cooperative college which can skill cooperative members who lack the principles and workings of a cooperative. The establishment of the cooperative college or university could boost the growth of cooperatives in South Africa.

9.3.7 Building a Strong Cooperative Financial Sector

The cooperative sector, because they are often small-scale, are constrained by lack of access to startup capital. In the Kenya context, the sector has been successful in building a strong financial cooperative sector that is anchored on member contributions since the early 1980s. The presence of strong financial cooperative institutions has ensured that startup cooperatives have access to capital which is not easily available from conventional financial institutions due to challenges such as the lack of collateral. In the South African context, cooperatives rely largely on the government for startup capital. Despite ongoing financial constraint, available evidence indicates that financial cooperatives are nearly non-existent. To address this challenge, the sector could learn from the success of self-help financial models such as *stokvel* that have been very successful in South Africa. Additionally, there is a need to emphasise member contributions as a key component of cooperatives in the training offered to cooperatives. This approach will ensure that those intent on forming cooperatives become fully informed and committed to making contributions to meeting the financial needs of their cooperatives. There is also a need to emphasise the diversification of the sector beyond agriculture being the primary focus to the provision of financial and non-financial services.

9.3.8 Extending Cooperatives Beyond the Realm of the Poor and the Vulnerable

While the country's approach towards cooperatives from the late 1990s to date all demonstrated a recognition of the value of cooperatives, they have largely focused on the activities of cooperatives among poor vulnerable communities. Cooperatives undoubtedly are significant actors in addressing the socio-economic conditions of vulnerable population but their scope of operation is limited to the small-scale sector. However, evidence from the review showed that cooperatives are not limited to poor and vulnerable communities. Cooperatives have and continue to participate in the global economy with positive outcomes for members and the countries where they operate. This view of cooperatives need to be emphasised in the South African context so as to make it appealing to members of the society who are not poor but are attracted by the principles and values of the cooperative movement. This could also be invaluable in attracting valuable financial capital which is a key challenge to the South African cooperative sector.

9.3.9 De-recialisation of the Cooperative Sector

The review of the history of the South African cooperative sector showed that cooperatives in colonial and Apartheid South Africa were established mainly to advance the collective interest of white South Africans. This race-based approach to cooperatives has not changed in post-Apartheid South Africa. Broadly speaking, government support, as espoused in its policies, strategies and Acts are located within the broader framework of the Broad-based Black Economic Empowerment programme.[1] Although the focus on black-owned cooperatives is noble as it seeks to address the inequalities of the past, it is a perpetuation of a dual cooperative system that characterised the previous regimes. Elimination of a dual cooperative system could be instrumental in advancing cooperation between black-owned and white-owned cooperatives.

9.4 Conclusion

Both the eradication of poverty and managing development that creates opportunities for the poor remain a challenge in developing countries such as South Africa. To address these challenges, the management of poverty should not only be for poverty eradication but also for mobilising economic freedom. Cooperatives have become an alternative to the traditional and conventional forms of economic systems. The cooperative model in South Africa was adopted as a form of social security intervention for poverty alleviation, employment creation and closing the gap of economic inequality. Despite the state interventions being regarded as noble interventions and support for the development and growth of cooperatives, co-ops are not completely free from political interference which in turn creates dependency on the state. This book has presented perspectives on the cooperative movement and highlighted some key lessons that could be applied to facilitate the growth of the country's cooperative sector. To become a successful sector, co-operators themselves should fight to defend the distinct identity upon which the co-op model is based. For the cooperatives to grow independently and interdependently, the government should provide a conducive and enabling environment aimed at supporting their initiatives.

---

[1]Black economic empowerment is one of the policy instruments used by the South African government to increase the share and participation of black South Africans in the formal economy. The operationalisation of this approach incudes a score-card to measure the transformation of corporate South Africa, and extending preferred options to businesses that meet government transformation targets. It also include preference for black South Africans in terms of government employment and procurement.

References

Cameron, R. (1996). *The Reconstruction and Development Programme*. J Theor Polit, Vol. 8 (2), 283–294.

Muthuma, W. (2011). *Economic cooperation in Kenyan credit cooperatives: Exploring the Rile of social capital and institutions*. Unpublished Ph.D. Johannesburg: Faculty of Commerce, Law and Management at the University of Witwatersrand.

Pivot, & Driving growth in KwaZulu-Natal. (2009). KwaZulu-Natal Official Publication of the ministry of Economic Development and Tourism. Pietermaritzburg: KZN-Ministry of Economic Development and Tourism.

Index

A.E. Okem (ed.), *Theoretical and Empirical Studies on Cooperatives*, SpringerBriefs in Geography, DOI 10.1007/978-3-319-34216-0

MIX
Papier aus verantwortungsvollen Quellen
Paper from responsible sources
FSC® C105338

If you have any concerns about our products,
you can contact us on
ProductSafety@springernature.com

In case Publisher is established outside the EU,
the EU authorized representative is:
Springer Nature Customer Service Center GmbH
Europaplatz 3, 69115 Heidelberg, Germany

Printed by Libri Plureos GmbH
in Hamburg, Germany